THE LATE VICTORIAN GOTHIC

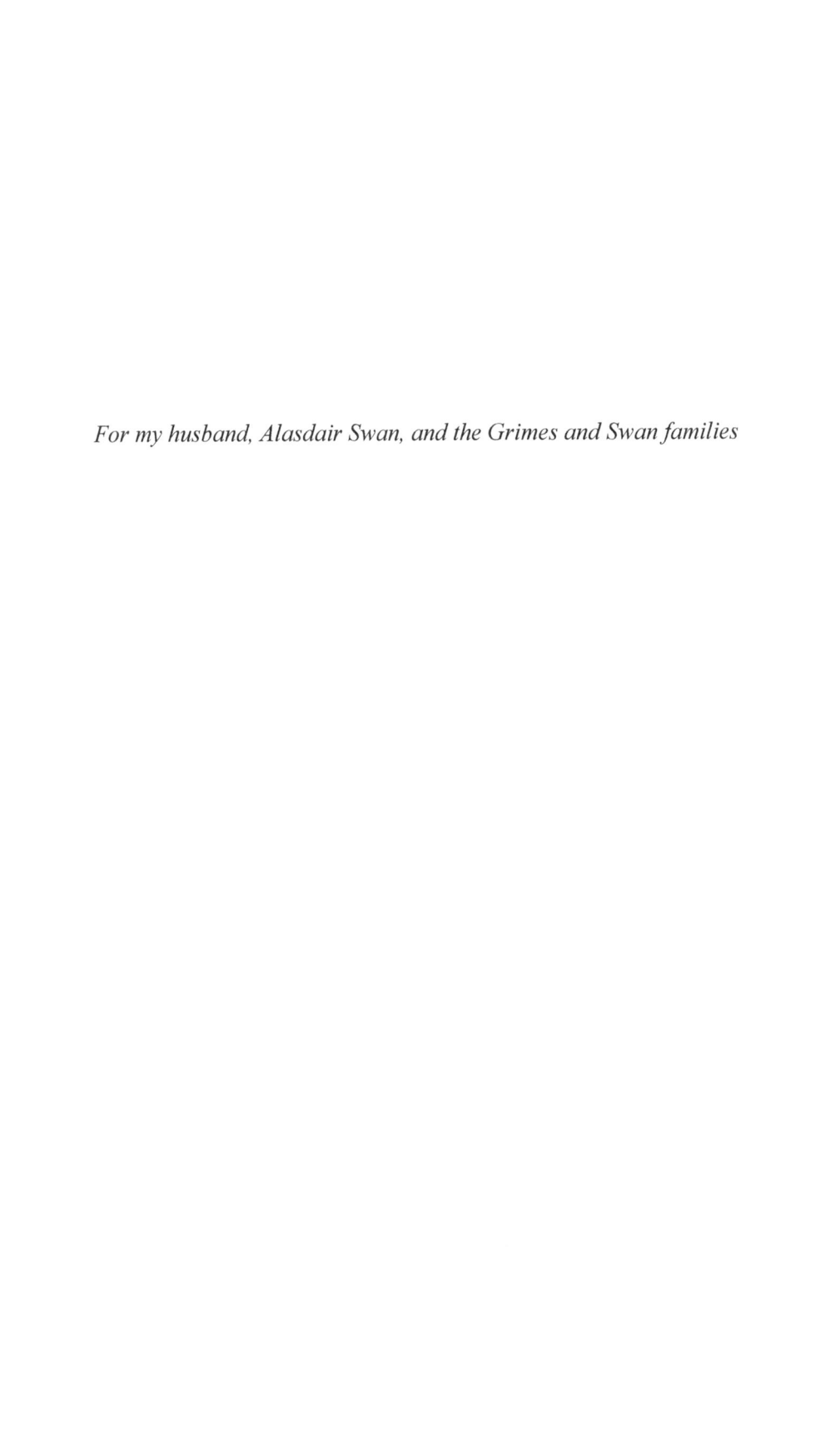

For my husband, Alasdair Swan, and the Grimes and Swan families

The Late Victorian Gothic

Mental Science, the Uncanny, and Scenes of Writing

HILARY GRIMES

Routledge
Taylor & Francis Group

LONDON AND NEW YORK

First published 2011 by Ashgate Publishing

2 Park Square, Milton Park, Abingdon, Oxon OX14 4RN
711 Third Avenue, New York, NY 10017, USA

Routledge is an imprint of the Taylor & Francis Group, an informa business

First issued in paperback 2016

British Library Cataloguing in Publication Data
Grimes, Hilary.
The late Victorian Gothic: mental science, the uncanny, and scenes of writing.
    1. Gothic revival (Literature) – Great Britain. 2. Gothic fiction (Literary genre), English
    – History and criticism. 3. Authorship – Social aspects – Great Britain – History –
    19th century. 4. Literature and technology – Great Britain – History – 19th century.
    5. Supernatural in literature. 6. Ghost stories, English – History and criticism.
    7. Literature and society – Great Britain – History – 19th century. 8. Gothic fiction
    (Literary genre), American – History and criticism.
    I. Title
    823'.087290908–dc22
Library of Congress Cataloging-in-Publication Data
Grimes, Hilary.
  The late Victorian Gothic: mental science, the uncanny, and scenes of writing / Hilary
  Grimes.
    p. cm.
  Includes bibliographical references and index.
    1. Gothic fiction (Literary genre), English—History and criticism. 2. English fiction—
  19th century—History and criticism. 3. Technology in literature. 4. Psychology in literature.
  5. Supernatural in literature. 6. Literature and technology—Great Britain—History—19th
  century. I. Title.
  PR878.T3G75 2011
  823'.087290908—dc22

                                                                          2011013777

ISBN 978-1-4094-2720-9 (hbk)
ISBN 978-1-138-26137-2 (pbk)

Contents

Acknowledgements   *vii*

Introduction   1

1   (Ghost)Writing Henry James: Mental Science, Spiritualism,
    and Uncanny Technologies of Writing at the Fin de Siècle   13

2   Sensitive to the Invisible: Photography and the Supernatural
    in the Holmes Stories, Arthur Conan Doyle's Spiritualism,
    and Francis Galton's Composite Portraits   37

3   Identities and Powers in Flux: Mesmerism, Hypnotism, and
    George Du Maurier's *Trilby*   61

4   Ghostwomen, Ghostwriting   83

5   Case Study: Vernon Lee, Aesthetics, and the Supernatural   111

6   Balancing on Supernatural Wires: The Figure of the
    New Woman Writer in Sarah Grand's *The Beth Book* and
    George Paston's *A Writer of Books*   137

Postscript   161

Bibliography   *163*
Index   *181*

Acknowledgements

This book was born out of research originally begun at the University of Glasgow. For their support and guidance, I would like to thank Alison Chapman, Andrew Radford, Richard Cronin, Nicola Trott, Glennis Byron, and Linda Dryden. Thanks must also go to the wonderful Churnjeet Mahn, Anna Barton, and Iain Kee Vaughan, who were there from the beginning, and also to all of my lovely Glasgow friends. I would also like to extend many thanks to Maggie Kilgour at McGill University for recognizing and nurturing my love of the Gothic, and to William Hughes of Bath Spa University for his kind words of encouragement and sound advice.

I would like to express my gratitude to Ann Donahue and Kathy Bond Borie at Ashgate for their understanding and ready assistance, as well as to the anonymous reader of my work, whose criticisms and comments were both instructive and insightful.

Early versions of Chapters 3 and 4 appeared in *Gothic Studies* and *The Victorian Newsletter* as 'Power in Flux: Mesmerism, Mesmeric Manuals and Du Maurier's *Trilby*,' *Gothic Studies*, 10 (November 2008), pp. 67–83, ISSN 1362–7937, and 'The Haunted Self: Visions of the Ghost and the Woman at the Fin-de-siècle,' *The Victorian Newsletter*, 107 (Spring 2005), pp. 1–4, ISSN 0042–5192. Thank you to the editors of both journals for granting me permission to publish the revised material here.

Many thanks indeed should go to both my own and my husband's relations – the Grimes and Swan families are truly great (and very patient). Most of all, I want to thank my husband, Alasdair Swan, for his endless cups of tea, unwavering support, and true love.

Introduction

The Lord thy God is the invisible stranger at the gate in the night, knocking. He is the mysterious life-suggestion, tapping for admission. And the wondrous Victorian age managed to fasten the door so tight, and light up the compound so brilliantly with electric light, that really, there *was* no outside, it was all in. The Unknown became a joke: is still a joke.[1]

Looking back on the Victorian period, D.H. Lawrence saw an age consumed by sterile faith in reason. In this passage he dismisses the Victorian interest in the 'Unknown' as a joke, using language evocative of the spiritualist séance: the 'knocking' on the door might be the rapping of spirits, eager to communicate messages from the other world. 'Spirit-rappings' marked the beginning of the spiritualist movement: in 1848, the Fox sisters of New York heard rappings in their house and reported that they were messages from the spirit world. The girls quickly became celebrities, and spiritualism was born, spreading rapidly from the United States to Britain. For Lawrence, however, spiritualism is rendered ridiculous in the harsh light of Victorian rationalism. But the deeper forces Lawrence trivializes are central to this book, which examines 'the mysterious life-suggestion, tapping for admission' in the late Victorian period and the ways in which writers and mental scientists of the fin de siècle were deeply conflicted between a desire to police the boundaries of science, identity, and the mind, and conversely, to experience the obscure thrill of the 'Unknown'. Although elements of the unknown like telepathy, spiritualism and spirits, mesmerism, and extrasensory perception threatened to compromise their rational borderlines, they were also intoxicating and inspiring, both dangerous and delightful.

While my principal focus is on Gothic literature written in the 1880s and 1890s, I also recognize and cut across strict boundaries, since texts like Henry James's 'The Private Life' (1892) and 'In the Cage' (1898), Rudyard Kipling's 'Wireless' (1902), Arthur Conan Doyle's Sherlock Holmes stories, George Du Maurier's *Trilby* (1894), ghost stories by women like Vernon Lee, and New Woman novels like Sarah Grand's *The Beth Book* (1897) and George Paston's *A Writer of Books* (1898) negotiate themes associated with both the Victorian and modernist periods, such as psychical research, mass marketing (especially literature), and new technologies. Indeed, I suggest that Gothic literature itself blurs boundaries, not only between literary periods, but also between genres. As Julian Wolfreys suggests, 'the gothic becomes truly haunting in that it can

---

[1]   D.H. Lawrence, *Kangaroo*, ed. by Bruce Steele, intro. by Macdonald Daly (London: Penguin, 1997), p. 285.

never be pinned down to a single identity'.[2] For Christine Berthin, 'it is in the nature of the Gothic to explode textual limits in order to include the hidden and the hinted at'.[3] While I argue for a more intimate Gothic which has the ability to infect or intermingle, rather than explode, I also suggest that rather than just including hidden themes, the Gothic is itself hidden within texts and genres which seem to resist this inclusion. I am particularly interested in texts which are not placed in the Gothic genre, but which nevertheless conceal Gothic themes, like horror about identity and doubling: New Woman fiction and Doyle's Sherlock Holmes stories, for example, both evoke and condemn supernatural themes. *The Late Victorian Gothic* explores how Gothic fiction, and the discourse surrounding mental science, spiritualism, and new technologies of the 1880s and 1890s, blend and bleed into one another, reflecting their subject matter, which also blurs together differing forces. For example, discussions of 'imponderable fluids' like ectoplasm blur distinctions between the material and the immaterial, since ectoplasm can be both visible and invisible, tangible and intangible. Furthermore, strict separations between the scientific and the supernatural begin to break down as scientific bodies like the Society for Psychical Research (SPR) increasingly investigate sites of the unknown like mesmerism and spirits, and spiritualists use scientific methodologies in order to authenticate, verify, and categorize the supernatural.

In my use of the term Gothic, I do not refer to the traditional tropes associated with the genre, like ruined castles, wicked villains, and helpless heroines. I do include what Andrew Smith calls 'certain persistent features' of the Gothic, such as 'the supernatural, and excess'; however, my aim is to argue for a new definition of the Gothic which is specific to the end of the nineteenth century.[4] This new definition follows along from Roger Luckhurst, who argues that '[t]he fin-de-siècle Gothic was fascinated by forms of psychic splitting, trance states, and telepathic intimacies. It adopted the language of the psychical researcher and relocated spooky phenomena within the quotidian spaces of English modernity; rather than the pre-modern landscapes of its eighteenth-century antecedent'.[5] Indeed, this study shows how the Gothic comprises moments in the text when characters come face to face with supernatural entities like ghosts, as well as when characters use trance states like mesmerism, hysteria, or dream states in order to access other powers within themselves. But while Luckhurst 'restore[s] some of the multivalencies to the fin-de-siècle Gothic, setting it in parallel with

---

[2]   Julian Wolfreys, *Victorian Hauntings: Spectrality, Gothic, the Uncanny and Literature* (Basingstoke: Palgrave Macmillan, 2002), p. 11.

[3]   Christine Berthin, *Gothic Hauntings: Melancholy Crypts and Textual Ghosts* (Basingstoke: Palgrave Macmillan, 2010), p. 2.

[4]   Andrew Smith, *Gothic Literature* (Edinburgh: Edinburgh University Press, 2007), p. 4.

[5]   Roger Luckhurst, *The Invention of Telepathy: 1870–1901* (Oxford: Oxford University Press, 2002), p. 182.

psychical research, as a fictional space for exploring reconfigurations of the self', this study argues that a Gothicism specific to the end of the century emerged which is concerned with instances of the uncanny within a text.[6] These uncanny moments are connected with the destabilization of the self during scenes of writing and in interactions with technologies of writing like the typewriter and the telegraph. How does technology make problematic/haunt the notion of agency in writing? Who has agency – the writer, the machine, or the medium? How does the notion of a spiritualist medium come to change conceptions of the working medium like the typist and telegraphist? I explore how, when notions of agency are suspended, authorship becomes uncanny, both a familiar practice and a deeply unfamiliar one, and suggest that this uncanniness is both terrifying and thrilling.

Indeed, this book addresses what is thrilling, particularly the *intoxicating* moments in the dialogue between science and the supernatural. To intoxicate is 'to poison', 'to stupefy, render unconscious or delirious, to deprive of the ordinary use of the sense of reason', 'to corrupt morally or spiritually', and also 'to render unsteady or delirious in mind or in feelings; to excite or exhilarate beyond self-control'.[7] Writers at the fin de siècle find moments of heightened perception both corrupting and exhilarating – when they experience multiplied consciousness they are experiencing a kind of ecstasy. While ecstasy is '[p]oetic frenzy or rapture', it is also a trance state which causes 'anxiety, astonishment, fear or passion'. This book suggests that writing at the fin de siècle is an ecstatic process, both entrancing and nightmarish. For example, Henry James and New Woman writers are both repelled by and drawn towards public acceptance and popularity. Furthermore, women writing ghost stories use horror as a powerful means of creating thrilling stories, but also of voicing concerns about women's political invisibility in the late nineteenth century. While much of the literature of the 1880s and 1890s is posited on a faith in rational progress as inevitable (Doyle's Sherlock Holmes, for example, seems to represent the triumph of logic), writers are still grappling with energies like telepathy and mesmeric fluids, which cannot be contained within the crisp contours of rational discourse. Finally, this book explores the ways in which fin-de-siècle writers are haunted by their attempts (and inevitable failures) to police the supernatural.

Recently, critics like Andrew Smith, Robert Mighall, and Kelly Hurley have also focused on late-nineteenth-century Gothic fiction in relation to Victorian science. Smith's examination of degeneration, sexology, and masculinity discusses the concerns of the medical community regarding representations of masculinity, but ignores how both scientists and fiction writers were trying to police identity (including gender constructs) and finding that identity could not be

---

6   Ibid., p. 185.

7   Oxford English Dictionary (OED). All subsequent definitions quoted in this book are taken from the OED, unless otherwise specified.

contained within the strict limits they imposed.[8] Like Smith, Mighall examines the exchange between fictional and non-fictional discourses. However, while Mighall focuses on the 'historical, geographical, environmental, and discursive factors which have played an important part in making Gothic representations credible at any given time', his decision to exclude psychological readings of the Gothic weakens his argument.[9] I contend that all of the factors he lists *and* psychology are crucial to a reading of the Victorian Gothic. While Hurley sees the importance of using both psychology and readings of medical and mental science texts in an understanding of fin-de-siècle literary culture, she is primarily concerned with examples of degeneration in Gothic texts rather than with haunted scenes of writing.[10]

Julian Wolfreys' *Victorian Hauntings: Spectrality, Gothic, the Uncanny and Literature* does consider what is haunting about writing. Although Wolfreys poses questions which underpin this book, like 'What does it mean to address the text as haunted? How do the ideas of haunting and spectrality change our understanding of particular texts and the notion of the text in general?', he is ultimately preoccupied with the haunting effect on the reader/audience rather than on the writer.[11] Problematically, Wolfreys suggests that 'the spectral effect … needs structure', implying that hauntings are dependent on materialism, whereas I argue that what concerned Victorian writers and mental scientists was the possibility that ghosts were both material bodies and immaterial presences.[12] Furthermore, Wolfreys situates his understanding of the haunted nature of writing in the mid-Victorian period, but uses his idea that ghosts disrupt texts to refer to Gothic fiction spanning from Dickens's *A Christmas Carol* (1843) to Stephen King's *The Shining* (1977). While ghosts and ghostliness do transcend distinctions, and while this book examines the blurring of boundaries between literary genres and scientific and supernatural discourses, I argue that the 1880s and 1890s represent a period of a new kind of Gothicism and are key to a discussion of haunted writing: during no other time were writers and mental scientists so anxious about the ways in which the supernatural was affecting the written word and the world of the mind.

Although these recent critics claim to offer an insight into neglected aspects of the Victorian Gothic, non-canonical writers are noticeably absent from their analyses. While Kelly Hurley brings some marginalized writers like Richard Marsh and Arthur Machen into her discussions, she and the other critics ignore women's writing entirely, focusing on canonical male Gothic writers like

---

[8]　　Andrew Smith, *Victorian Demons: Medicine, Masculinity and the Gothic at the Fin-de-Siècle* (Manchester: Manchester University Press, 2004), p. 5.

[9]　　Robert Mighall, *A Geography of Victorian Gothic Fiction: Mapping History's Nightmares* (Oxford: Oxford University Press, 1999), p. xiv.

[10]　　Kelly Hurley, *The Gothic Body: Sexuality, Materialism, and Degeneration at the Fin de Siècle* (Cambridge: Cambridge University Press, 1997).

[11]　　Wolfreys, p. ix.

[12]　　Ibid., p. 5.

Charles Dickens, Oscar Wilde, Robert Louis Stevenson, and Bram Stoker. My aim is to discuss well-known authors like Henry James and Arthur Conan Doyle alongside lesser known, but significant writers like Anna Bonus Kingsford and Mary Louisa Molesworth, not only to bring neglected voices to the forefront of literary study, but to demonstrate that these voices articulate nagging anxieties about writing in the late nineteenth century. Indeed, a recovery of women writers is crucial to my study.[13] I devote my last three chapters to women's writing because examining it is essential to an understanding of the late Victorian Gothic, spiritualism, psychical research, and nineteenth-century discursive technologies. In doing so, I am following the example of recent studies like Jill Galvan's *The Sympathetic Medium: Feminine Channeling, the Occult and Communication Technologies 1859–1919* and Marlene Tromp's *Altered States: Sex, Nation, Drugs, and Self-Transformation in Victorian Spiritualism*, which have placed discussions of women and spiritualism at the centre of Victorian studies. These studies articulate spiritualism's 'social power', particularly its negotiation with key anxieties of the period – in Galvan's case how mediumship (referring to both the activity of the woman at a séance and her use of emerging communication technologies) was connected to concerns about privacy of information,[14] and in Tromp's case 'the ways in which [spiritualism] may have facilitated cultural change', particularly because it 'participated, with many other social movements, in the transformation of culture-wide notions of gender, race, and social propriety'.[15] While my focus in the latter half of this book is on women writers rather than on mediums, I am uncovering the ways in which women writers, like men writers of the period, were in dialogue not only with spiritualism and technologies of writing, but also with contemporary theories of mental science. Even more than their male contemporaries, though, and because of their desire to achieve serious literary approval rather than be associated with popular forms like the ghost story (and the symbolic threat of becoming ghostly in the canon), women writers were both repulsed and captivated by the desire to write about haunting.

Interest in terms and preoccupations like mesmerism, telepathy, spiritualism, and double consciousness has been growing in Victorian studies. Pamela Thurschwell's *Literature, Technology and Magical Thinking, 1880–1920* and Roger Luckhurst's *The Invention of Telepathy: 1870–1901* have been instrumental not only in reviving these neglected aspects of the period, but also

---

[13] Although Diana Basham's *The Trial of Woman: Feminism and the Occult Sciences in Victorian Literature and Society* (London: Macmillan, 1992) has made strides towards examining Victorian women's writing and its links to the occult, the work focuses on the mystical associations of women and menstruation and ignores other important aspects of women's fiction – in particular, the uncanny scenes of writing that figure women's relationship to fiction itself.

[14] Jill Galvan, *The Sympathetic Medium: Feminine Channeling, the Occult, and Communication Technologies, 1859–1919* (Ithaca: Cornell University Press, 2010).

[15] Marlene Tromp, *Altered States: Sex, Nation, Drugs, and Self-Transformation in Victorian Spiritualism* (Albany: State University of New York Press, 2006), p. 5.

in demonstrating that a multidisciplinary approach – examining contemporary fiction alongside emerging theories of mental science and psychical research, for example – offers the most salient means of understanding the late Victorian period. Thurschwell's study focuses on the eroticism that interactions with the occult and technology could evoke, as well as the ways in which late-nineteenth-century fiction and mental science served as the precursors to psychoanalysis. However, she does not examine in sufficient detail how new technologies changed and made uncanny both the writer and the act of writing.[16] Luckhurst's excellent study of telepathy and the Gothic is influential on my own thinking, and in some ways this book begins with and expands out of his argument that at the fin de siècle:

> the supernatural did not simply conjure monsters, but figured emergent conceptions of psychical life. With a *spectralized* subject, late Victorian fiction developed the uncanny discoveries of hypnotists and psychical researchers, and pursued the possibility that (as William Stead put it) 'each of us has a ghost inside him'.[17]

This book explores the ghost within, not only in relation to contemporary studies of the mind, spiritualist practice, and writers' negotiation with new technologies, but also, and especially, in connection with scenes of writing and anxieties about the dangerous similarities between science and the supernatural in the 1880s and 1890s. In doing so, *The Late Victorian Gothic* makes a dynamic intervention in Victorian studies, articulating a Gothic vision of the end of the century. The texts and authors I have chosen are intended to be representative of this Gothicism. I examine writers from a wide variety of backgrounds, disciplines, and genres – aesthetic writers like Vernon Lee, popular writers like George Du Maurier, serious canonical writers like Henry James, spiritualists, and mental scientists – to demonstrate how far-reaching concerns about writing were in all aspects of the late Victorian period. Furthermore, I discuss texts like women's ghost stories, mesmeric handbooks, mental science and medical texts, Doyle's spiritualist writings, and forgotten New Woman novels to show that these peripheral works are actually central expressions of fin-de-siècle anxieties.

Haunting this book is the notion of the supernatural and the uncanny. Nicholas Royle argues that 'the uncanny is not simply synonymous with the supernatural …, but is more accurately *suggestive* of – "associated with", or "seeming" to have a basis in – the supernatural'.[18] Indeed, 'uncanny' and 'supernatural' are two distinct terms, but the separation between them is more profound than Royle implies. Whereas the supernatural relates to the external, to disturbances in the exterior world, the uncanny is psychological, representing disturbances in the internal

---

[16]   Pamela Thurschwell, *Literature, Technology and Magical Thinking, 1880–1920* (Cambridge: Cambridge University Press, 2001), p. 2.

[17]   Luckhurst, *The Invention of Telepathy*, p. 213.

[18]   Nicholas Royle, *The Uncanny: An Introduction* (Manchester: Manchester University Press, 2003), p. 10.

body, or mind: in other words, the supernatural is a cause and the uncanny an effect.

'Supernatural' meant many things in the nineteenth century, which, as Nicola Bown, Carolyn Burdett, and Pamela Thurschwell have argued, was part of its appeal. They define it as 'slipper[y]' and 'resistan[t] to definition', suggesting it had a 'protean quality of being a cause, a place, a kind of being, a realm, a possibility, a new form of nature, [and] a hope for the future'.[19] Richard Noakes argues that the Victorian interest in spiritualism was key in redefining ways of thinking about the supernatural:

> Victorian investigators of Spiritualism believed that the erratic phenomena of the séance could be reduced to natural laws and that their enterprises could thereby gain scientific credibility. However, this was a difficult goal to achieve since many critics of Spiritualism questioned the very possibility of a naturalistic approach to phenomena that were ostensibly beyond nature, and actively defined the natural sciences in opposition to Spiritualism.[20]

Noakes reveals how supernatural events in spiritualism, such as full-form materializations, automatic writing, and table-rappings, could be construed not only as elements outside of nature, but as new kinds or manifestations of nature that had simply been overlooked or misconstrued in the past. Different interpretations of the supernatural could allow spiritualists to believe in other-worldly presences and sceptics like William Benjamin Carpenter to argue that the phenomena at séances were a result of 'the laws of mental action'.[21] The psychic investigator William Crookes, in fierce and public opposition to Carpenter, believed that spirit manifestations could be explained if new natural laws could be discovered. *The Late Victorian Gothic* interrogates all of these rich contemporary definitions, as well as the modern definition: '[t]hat is above nature; belonging to a higher realm or system than that of nature; transcending the powers of the ordinary course of nature', in order to discuss a supernaturalism that is a 'mysterious life-suggestion', a spectral concept haunting the writing of the 1880s and 1890s.[22]

My use of the word 'uncanny' is heavily influenced by Freud's essay 'The Uncanny' (1919), and I intend the term to resonate with all of his layered definitions of the word.[23] For example, Freud suggests that 'on the one hand [uncanny] means what is familiar and agreeable, and on the other, what is concealed and

---

[19]   Nicola Bown, Carolyn Burdett, and Pamela Thurschwell, 'Introduction', in *The Victorian Supernatural*, ed. by Nicola Bown, Carolyn Burdett, and Pamela Thurschwell (Cambridge: Cambridge University Press, 2004), pp. 1–19 (p. 8).

[20]   Richard Noakes, 'Spiritualism, Science and the Supernatural in Mid-Victorian Britain', in Bown, Burdett, and Thurschwell, eds, *Victorian Supernatural*, pp. 23–43 (p. 24).

[21]   Ibid., p. 32.

[22]   D.H. Lawrence, p. 258.

[23]   See Sigmund Freud, 'The Uncanny', in *The Standard Edition of the Complete Psychological Works of Sigmund Freud*, ed. and trans. by James Strachey, 24 vols

kept out of sight' (pp. 224–5). Freud lists 'things, persons, impressions, events and situations which are able to arouse in us a feeling of the uncanny' (p. 226), such as 'doubling, dividing and interchanging of the self' (p. 234), the 'factor of the repetition of the same thing' (p. 236), 'death', 'spirits', 'ghosts' (p. 241), 'magic and sorcery', [and] 'the omnipotence of thoughts' (p. 243). Although Freud is tentative about including the 'intellectual uncertainty whether an object is alive or not, and when an inanimate object becomes too much like an animate one' as 'a condition for awakening uncanny feelings' (p. 233), this 'condition' is a valuable means of discussing uncanny technologies at the end of the century, particularly in relation to the emergence of women in the workforce: were automatic writers (a term referring to spiritualist mediums and typists) becoming uncanny in their interactions with technology? Were they becoming mechanized? Could new technologies reproduce human actions better than people could?

In my understanding of the uncanny, I also draw on some of Nicholas Royle's definitions, which are particularly useful to a study on blurred distinctions between sites of power and identities. Royle suggests that the uncanny 'is ghostly',[24] that it 'can involve a feeling of something beautiful but at the same time frightening, as in the figure of telepathy or doubling', and that it 'has to do with a strangeness of framing and borders, an experience of liminality'.[25] This book is very much concerned with this type of uncanny – it explores the moments in which writers conjure up and then exorcise the supernatural in their fiction, playing out their fears and desires about the boundaries separating science and the unknown. Finally, I suggest, as Freud, Royle, and others have done, that the uncanny has everything to do with the writing process. In a discussion of writers producing uncanny effects, Freud argues that

> [the writer] can even increase his effect and multiply it far beyond what could happen in reality, by bringing about events which never or very rarely happen in fact. In doing this he is in a sense betraying us to the superstitiousness which we have ostensibly surmounted; he deceives us by promising to give us the sober truth, and then after all overstepping it. (p. 250)

Here, Freud expresses the anxiety that writers in the late nineteenth century felt about the act of writing. Writers promised to give 'the sober truth', and to have absolute control of their body of work, both during the writing process and once it was published and became part of the public market. But uncannily, there were always outside energies 'tapping for admission',[26] and the work was never entirely as tidy, contained, and uninfluenced as writers would have liked. Although Freud insists that writers betray and deceive the reader, it is really the writers' own

---

(London: Hogarth Press, 1953–1974), XVII, pp. 219–56. Subsequent references to this edition are given after quotations in the text.

[24]   Royle, *Uncanny*, p. 1.

[25]   Ibid., p. 2.

[26]   D.H. Lawrence, p. 285.

intent that is deceptive: despite being anxious about the supernatural, fin-de-siècle writers secretly desired to 'overstep' their self-imposed boundaries (p. 250).

In Chapter 1, I examine the ways in which mental scientists and writers of the 1880s and 1890s were striving to maintain 'academic neatness', both in studies of the mind and in writing. Henry James's 'The Private Life' shows that writing always escapes perfect order: rather than offering unity to the author, scenes of writing actually split the writing self between public and private personae, and the author is haunted by the figure of his own celebrity. Furthermore, a discussion of James's typist, Theodora Bosanquet, shows that the typewriter increased James's uneasiness about the act of writing. As a medium for James's message, Bosanquet gained an agency in the act of transcribing which suggested the site of authorship itself was unstable. Finally, in a close reading of James's 'In the Cage' and Rudyard Kipling's 'Wireless', this chapter examines the uncanny ways in which new technologies fragmented identity. Despite attempts to police the boundaries of the self, new technologies and new understandings of the mind only increased the haunting nature of writing.

In Chapter 2, I continue my discussion of the uncanny effects of technology, examining late Victorian perceptions of photography and Arthur Conan Doyle's complicated relationship with the art. An examination of Francis Galton's composite portraits, and of spirit photography, suggests that at the end of the century, photography came to symbolize the uneasy dynamic between science and supernaturalism that this book articulates: the photograph was representative of both unassailable scientific evidence *and* the late Victorian fixation with the ghostly. This chapter discusses Doyle's own photographic work, and how his interest shifted from the scientific process of developing pictures in the darkroom to his fascination with spirit photography and spiritualism. Like other spiritualists and mental scientists of the period, Doyle wanted, impossibly, to scientifically contain and control the supernatural. This chapter connects conflicting ideas about the photograph and about Doyle himself in order to re-examine conceptions about Sherlock Holmes, who, like the plate of the camera, is sensitive to elements normally invisible to rational perception, such as the supernatural. Surprisingly, not only do his deductive abilities come from spiritual insight, but he also gains heightened perception from mesmeric and hypnotic trance states.

Chapter 3 focuses on mesmerism and hypnotism, exploring the exchange of powers during trance states in the context of the fin de siècle and analyzing their treatment in the fields of mental science, spiritualism, and the Gothic literature of the 1880s and 1890s. I examine the ways in which mental scientists in the late Victorian period attempted to introduce hypnotism as a valid scientific field of study, and how they were nevertheless drawn to the supernatural possibilities of mesmerism. Attempts by mental scientists to police the boundaries between the supposedly scientific hypnotism and supernatural mesmerism only demonstrated how uncannily close these discourses were, which blurred into one another and made each other indistinct. Furthermore, a re-examination of mesmerism and hypnosis manuals demonstrates that these trance states were also about blurring powers and identities, something that was deeply threatening for the late Victorians. Startlingly, the notion of the all-powerful mesmerist/hypnotist is not supported by the manuals, and instead, mesmerism and hypnotism are marked by shared sites of power. In a close reading of George Du Maurier's *Trilby*, I show that Trilby and Svengali share power in scenes of mesmerism and hypnotism, a possibility more horrifying to the other characters in the text than the notion of an all-powerful Svengali.

Trance states are crucial to a discussion of women's writing in Chapters 4 and 6. For example, Chapter 4 examines how women in ghost stories see ghosts in altered states of perception like hysteria, hypnosis, and dream states and identify with them. Furthermore, women ghost-story writers use the ghost as a symbol for their political invisibility. While mental science was attempting to police the mind, medical science was attempting, through the discourse on hysteria, to police the female mind and body. Through a close reading of W.T. Stead's *Real Ghost Stories* (1891) and a case study of the SPR's investigation into a haunted house, I argue that theories of the unconscious mind as a haunted site were symbolic of the ghostly role of women politically, socially, and legally. In a discussion of non-canonical women ghost-story writers and their work, I show how women were both haunted and empowered by writing about ghosts.

In Chapter 5 I turn to a case study of Vernon Lee. Lee is particularly significant in a study that explores the ways in which mental scientists, writers, and psychical researchers of the 1880s and 1890s were both drawn to and radically anxious about discussing the supernatural, because her writing is a continuous struggle between science and supernaturalism. This chapter focuses on the links between aestheticism and ghostliness that Lee rejected and yet evoked, and demonstrates that Lee's attempts to dissociate aesthetics and the supernatural only pushed them more closely together. While Lee's psychological aesthetics were designed to be an empirical study of the body's reactions to art, her methods closely resembled those of the spiritualists and the members of the SPR in their interest in and attempts to record physiological reactions of the body to ghosts. Furthermore, while she insists on the fundamental materialism of aesthetics, her art is instead haunted by its inability to produce that materialism. This chapter also explores themes of writing and the uncanny, and how, in Lee, genre itself becomes haunted as her fiction and non-fiction texts become increasingly indistinct. Finally, this chapter places Lee within the context of the fin de siècle, demonstrating that although she is usually considered to be an exception to the literary trends and social concerns of the period, her interest in the supernatural and mental science make her representative of the late Victorian Gothic I am articulating.

Finally, in Chapter 6, I return to a discussion of the ways in which women writers use altered states of perception creatively. In a discussion of New Woman texts like Sarah Grand's *The Beth Book* and George Paston's *A Writer of Books*, I show that for women writers like Beth and Cosima, heightened perception is the site of ecstatic inspiration: writing with multiplied perception is both rapturous and debilitating. This chapter also returns to a discussion of the uncanny effects of technology, and in a close reading of Grant Allen's *The Type-Writer Girl* (1897), suggests that in New Woman texts, technology like the typewriter actually grants women like Juliet the freedom to choose whether they will engage with machines analytically or mechanically. This chapter is about blurred distinctions between automatic and active writing, between exhilaration and despair, and between realist and supernatural fiction.

From Henry James's anxieties about controlling the writing self and sites of agency to Doyle's concerns about capturing the immaterial on the sensitive plate of the camera, and from mental scientists' assurances that they could exorcise mesmerism from the occult by using occult language and themes to the attempts by mental scientists to contain the female mind and body within the same discourse as haunted states of mind, *The Late Victorian Gothic* is about the desire and failure to rigidly control identities, minds, and scenes of writing which preoccupied writers, mental scientists, and spiritualists at the fin de siècle. Finally, from Lee's insistence on making a strict separation between science and the supernatural, which her ghost stories would always overstep, to New Woman fiction writers who revelled in and were repulsed by the supernatural in their political texts, writers of the 1880s and 1890s were always haunted by and haunting scenes of writing.

Chapter 1
(Ghost)Writing Henry James: Mental Science, Spiritualism, and Uncanny Technologies of Writing at the Fin de Siècle

Introduction

> [T]he human mind [...] was largely an abstraction. Its normal adult traits were recognized. A sort of sunlit terrace was exhibited on which it took exercise. But where that terrace stopped, the mind stopped. [...] But of late years the terrace has been overrun by romantic improvers, and to pass to their work is like going from classic to Gothic architecture, where few outlines are pure and where uncouth forms lurk in the shadows. A mass of mental phenomena are now seen in the shrubbery beyond the parapet. Fantastic, ignoble, hardly human, or frankly non-human are some of the new candidates for psychological description. The menagerie and the madhouse, the nursery, the prison, and the hospital, have been made to deliver up their material. The world of mind is shown as something infinitely more complex than was suspected; and whatever beauties it may still possess, it has lost at any rate the beauty of academic neatness.[1]

In his commemoration lecture on F.W.H. Myers, the distinguished researcher in mental science and founding member and past president of the SPR, William James maintains that the mind had long been studied in a factual way, as orderly and linear. James offers the metaphor of a well-groomed terrace and asserts that, with the formation of the SPR in 1882 and especially with the influence of Myers, studies in psychology at the end of the century turned not only to what is on the terrace, but to what lies beyond it: what exists beyond and beneath the charted depths of the mind?

William James, Henry James's brother, was a philosopher, a psychologist, and president of the SPR from 1894 to 1895. The fact that William, one of the most widely read psychologists of the nineteenth century, writes about the transformation of the terrace from garden to jungle using literary language is suggestive of the ways in which the literature and science at the end of the Victorian period were feeding into one another. That the terrace is overrun by 'romantic improvers' implies that the field of mental science has been invaded by novelists whose imaginative 'material' seeps into psychological studies and blurs distinctions between scientific fact and fiction. His account of the classical studies into the mind implies that, as in the classical period, earlier researches

---

[1]   William James, 'Frederic Myers's Service to Psychology', in *A William James Reader*, ed. by Gay Wilson Allen (Boston: Houghton Mifflin, 1971), pp. 155–64 (p. 156).

emphasized established analysis and 'academic neatness'. Researches by the fin de siècle, however, were, like the Gothic literature he evokes, revelling in the 'menagerie and the madhouse'.[2] Although the 'beauty of academic neatness' is lost, James suggests there is something delightful, if frightening, in an inchoate version of the mind.

Significantly, the emergence of James's 'Gothic architecture' of mental science at the end of the century coincides with a revival of Gothic literature – *Strange Case of Jekyll and Hyde* (1886), *The Picture of Dorian Gray* (1890–1891), and *Dracula* (1897) are just some of the more popular examples of this trend. The Gothic studies of the mind and Gothic literature were not just overlapping, however, but were also in negotiation with one another, informing and inspiring each other. Stevenson's *Jekyll and Hyde*, for example, had been influenced by F.W.H. Myers's theories on subliminal consciousness. Upon reading the novel, Myers had written Stevenson a letter expressing his admiration, but also giving advice on how Stevenson could make 'medical and psychological improvements'.[3] Mental scientists and writers of Gothic fiction were also negotiating their shared anxieties about the stability of identity. They, and as this chapter will show, in particular Henry James, were haunted by the possibility that selfhood was itself a collection of illusional material.

While William James suggests that the 'Gothic architecture' of mental science can no longer be contained within the conventional rubrics of psychological research, his brother Henry is anxious about the potential loss of 'academic neatness' in his writing. Henry employs an architectural metaphor in his discussion of the 'house of fiction':

> The house of fiction has in short not one window, but a million […] every one of which has been pierced, or is still pierceable […] by the need of the individual will. These apertures, of dissimilar shape and size, hang so, all together, over the human scene that we might have expected of them a greater sameness of report than we find. They are but windows at the best, mere holes in a dead wall, disconnected, perched aloft; they are not hinged doors opening straight upon life. But they have this mark of their own that at each of them stands

[2]   Fin-de-siècle Gothic fiction deals with themes like menageries, madhouses, and monsters. For example, part of Bram Stoker's *Dracula* (1897) is set in an insane asylum, where Dracula visits the patient Renfield. Richard Marsh's *The Beetle* (1897) is about a terrifying creature which sometimes takes the form of a monstrous beetle, and which terrorizes the politician Paul Lessingham.

[3]   Roger Luckhurst, *The Invention of Telepathy: 1870–1901* (Oxford: Oxford University Press, 2002), p. 192. Luckhurst argues that in the late nineteenth century, 'psychology and Gothic fiction informed each other' (p. 193). Robert Mighall also theorizes a direct interchange of information between novelists and scientists of the period, arguing that scientists derived as many of their ideas about human identity from fiction as they did from scientific investigation. See *A Geography of Victorian Gothic Fiction: Mapping History's Nightmares* (Oxford: Oxford University Press, 1999). I suggest that Gothic and scientific writing influenced and *inspired* one another.

a figure with a pair of eyes, or at least with a field-glass, which forms, again and again, for observation, a unique instrument, insuring to the person making use of it an impression distinct from every other. He and his neighbours are watching the same show, but one seeing more where the other sees less, one seeing black where the other sees white, one seeing big where the other sees small, one seeing coarse where the other sees fine. And so on, and so on; there is fortunately no saying on what, for the particular pair of eyes, the window may *not* open [...]. The spreading field, the human scene, is the 'choice of subject'; the pierced aperture ... is the 'literary form'; but they are [...] as nothing without the posted presence of the watcher – without, in other words, the consciousness of the artist. Tell me what the artist is, and I will tell you he has *been* conscious.[4]

Henry James's 'house of fiction' is a haunted house, with ghostly guests lurking at every window. The structure also evokes William's overrun terrace, a Gothic mansion that is architecturally incoherent, unwieldy, and rambling, filled with shadows, spectral presences, and incongruous parts, straining away from the 'academic neatness' of a conventional house. The metaphor of the 'house of fiction' suggests that fiction itself cannot be regulated and has little unity, but rather a million vantage points, and that even the act of writing is a vexed, haunted endeavour. Indeed, the house of fiction seems to be haunted by James himself, whose consciousness is the watcher in the windows.

James's haunted 'house of fiction' and his brother's 'Gothic architecture' are apt symbols for anxieties about identity and writing at the fin de siècle.[5] Writers like Henry James and Vernon Lee (whom I will discuss in depth in Chapter 5) and mental scientists like Myers wanted to confine their work to tidy blueprints, but were instead operating within an unchartable Gothic landscape, overrun with ghosts. The connection at the end of the century between writing and ghostliness, and the close ties between science and supernaturalism, rise out of the contemporary

---

[4]   Henry James, 'The Portrait of a Lady', in *Literary Criticism: French Writers; Other European Writers; The Prefaces to the New York Edition*, ed. by Leon Edel (New York: Library of America, 1984), pp. 1070–85 (p. 1075). In a discussion of James's 'house of fiction' metaphor, Leon Edel outlines James's anxiety about writing and criticizing novels: 'One draws boundaries round novels at one's peril. [...] the critic of fiction is asked to accommodate himself not only to the novel's rhetoric but to the precise *field of vision* of the novelist, not to speak of his instruments of vision'. From Henry James, *The House of Fiction: Essays on the Novel*, ed. by Leon Edel (London: Hart-Davis, 1957), p. 11. While Edel explores James's problems as a literary critic, I suggest his concerns about not being able to create boundaries within his writing are linked to his anxieties about the ghostly and uncanny nature of writing itself.

[5]   See also critics like Tim Armstrong, Michel Foucault, and Bette London, who have discussed the ways in which nineteenth-century authorship was a problematic concept, difficult to define or limit. Tim Armstrong, *Modernism, Technology and the Body: A Cultural Study* (Cambridge: Cambridge University Press, 1998); Michel Foucault, *The Archaeology of Knowledge*, trans. by A.M. Sheridan Smith (London: Tavistock, 1972); Bette London, *Writing Double: Women's Literary Partnerships* (Ithaca: Cornell University Press, 1999).

discourse on mental science, spiritualism, and discursive technologies. Automatic writers at a spiritualist séance, for example, would transcribe the words of the dead, succumbing to a trance-like state and writing whatever the spirits relayed.[6] But they also raised questions about the nature of agency in scenes of writing: Should ghosts be given credit for published works, or the mediums who transcribed them?[7] How did automatic writing change the notion of influence and control for authors like James and Lee?

Studies by the SPR on the activities of mediums, and particularly automatic writing, implied that it was the subconscious mind, rather than external entities, which was responsible for the spiritualistic phenomena.[8] Myers wrote about the subject in 'Automatic Writing or the Rationale of the Planchette' (1885), arguing that automatic writing is the result of unconscious cerebration: writing could be physical evidence for activity in the brain of which a person is completely unaware.[9] In the field of mental science, studies on a 'secondary self' within indicated that it was not only the séance that was haunted but also, disturbingly, the mind.

Automatic writing in the Victorian period referred not only to spiritualism and the emerging science of mind, but also to discursive technologies like the phonograph (1877), the typewriter (c. 1868), and the telegraph machine (1838). Significantly, manifestations during spiritualist séances would follow new innovations in technology, suggesting that contrivances designed for clear and easy communication had been appropriated for the mediated form of communication between this world and the next. For example, when the Morse code was introduced in the mid- to late 1830s, spirits rapped in Morse, and with the invention of the telephone, disembodied voices at séances became the vogue.[10]

---

[6]   For more on the history of spiritualism at the fin de siècle, see, for example, Janet Oppenheim, *The Other World: Spiritualism and Psychical Research in England, 1850–1914* (Cambridge: Cambridge University Press, 1985), and Alex Owen, *The Darkened Room: Women, Power and Spiritualism in Late Victorian England* (London: Virago, 1989).

[7]   See Helen Sword, *Ghostwriting Modernism* (Ithaca: Cornell University Press, 2002). Sword suggests that assigning agency was problematic even in library catalogues, which often list books published by mediums under the author heading of the mediums' spirit guide (pp. 28–9). In *Spiritualism and Women's Writing: From the Fin de Siècle to the Neo-Victorian* (Basingstoke: Palgrave Macmillan, 2009), Tatiana Kontou also discusses how mediums disrupt agency, although her focus is on the ways in which Neo-Victorian writers adopt the role of the spiritualist medium. For more on influence in writing, see also Robert Douglas-Fairhurst, *Victorian Afterlives: The Shaping of Influence in Nineteenth-Century Literature* (Oxford: Oxford University Press, 2002). Douglas-Fairhurst suggests that 'detecting another hand in one's writing raises awkward questions about what it means for one's self to be influenced and therefore not one self only' (p. 5).

[8]   Oppenheim, pp. 123–33.

[9]   F.W.H. Myers, 'Automatic Writing or the Rationale of the Planchette', *The Contemporary Review*, 47 (1885), 233–49.

[10]   Steven Connor, *Dumbstruck: A Cultural History of Ventriloquism* (Oxford: Oxford University Press, 2000), pp. 364–86.

While Lisa Gitelman's *Scripts, Grooves and Writing Machines* discusses the ways in which writing machines radically refashion writing itself, I want to examine how technologies of writing, in particular the typewriter and the telegraph, alter the content of what is written or the mindset of the writer.[11] Friedrich Kittler's argument that identity fragments because of writing technologies is useful here – what happens to James, for example, when he begins to use the typewriter to create his work (and furthermore, when he employs an amanuensis to do the typing)?[12]

In an examination of Henry James's 'The Private Life', this chapter addresses the faltering attempts of writers and mental science alike to police the boundaries of identity and map the mind. Close readings of James's 'In the Cage' and Rudyard Kipling's 'Wireless' suggest that the technological innovations that were themselves writing machines contributed to the development of an uncannily multiplied selfhood. This chapter also examines Henry James's desire and inability to find 'academic neatness' within his 'house of fiction', looking in particular at James's anxieties about his writing when it reaches the reading public. Who owns the 'house of fiction'? How does celebrity become a haunting figure in James's short stories? For James, in spite of himself, all scenes of writing are haunted ones.

Authorship and the Uncanny: James's 'The Private Life'

Henry James's 'The Private Life' disturbingly indicates that authorship demands that the writer must suffer a separation into two distinct but dependent identities, the public self and the private self.[13] Perhaps James wants to imply that a public life is perversely necessary in order to preserve the private one: does the public self ironically make the private self more significant? In 'The Private Life' writing has an uncanny effect on selfhood. In 'The Uncanny', Freud suggests that the uncanny is the fear we feel when something 'is familiar and agreeable [and] concealed and kept out of sight' (pp. 224–5). It is within this conception of the uncanny that we find James's Clare Vawdrey, the writer who is almost shockingly consistent, who 'never talked about himself, and liked one subject [...] precisely as much as another' (p. 191), and whose 'opinions were sound and second-rate' (p. 192); he is, in effect, boring.[14] And yet Vawdrey's creative outpourings are stunningly beautiful and exciting: he is 'the greatest [...] of our literary glories' (p. 189). That he is dull in public and a brilliant writer in private is not, however,

[11] Lisa Gitelman, *Scripts, Grooves, and Writing Machines: Representing Technology in the Edison Era* (Stanford: Stanford University Press, 1999), pp. 1–20.

[12] Friedrich Kittler, *Discourse Networks, 1800/1900*, trans. by Michael Metteer (Stanford: Stanford University Press, c. 1990).

[13] Henry James, 'The Private Life', in *The Complete Tales of Henry James*, ed. by Leon Edel, 10 vols (London: Hart-Davis, 1962–1964), VIII, pp. 189–227. Subsequent references to this edition are given after quotations in the text.

[14] Clare Vawdrey is based on the poet Robert Browning. James was puzzled by the fact that Browning could write so well and yet socialize so badly.

what makes him so singular; what makes him so singular is that Vawdrey is, in fact, double. He has a bland, reliable, public self and a private genius self. The division between the writer and the public self seems an unequal one, however, since the writer seems to have consumed the personality of the public self.

'The Private Life' reveals that the writer is haunted by a ghostly other self who is both himself and someone else, making authorship within the text slippery: Who is holding the pen and controlling the words that flow from it? Who owns the work when it becomes part of the literary marketplace? In asking these questions, James's story is negotiating with contemporary concerns about authorship, particularly in relation to automatic writing – if writers acted as mediums for ghosts, how did that alter writing itself and the makeup of identity? The scene in which the narrator finds Vawdrey writing alone in the dark evokes the spiritualist séances in which the medium would sit transcribing while spirits manifested themselves outside. Like a medium taking down the words of ghosts, Vawdrey is in a 'fit' of 'abstraction', so possessed by the act of writing that he devotes all of himself to it: in turn, his public self becomes a materialized spirit, only the ghost of a personality (p. 206). While in the field of mental science automatic writing highlighted that the mind could no longer be contained within the tidy boundaries of 'academic neatness', in 'The Private Life' creative writing also denies writers this precision and clarity. Instead, the act of writing transforms the writer into an unfathomable, ghostly, and multiplied presence.

In 'The Private Life' writing destabilizes boundaries of identity to such an extent that selfhood itself is negotiated by literary language, particularly in the case of Lord Mellifont's public identity. His dress, manners, and all his actions are the 'topic' of discussion, the 'subject' (p. 196) gracing everyone's lips, and they set the 'tone' (p. 197) for every occasion. In fact, he seems to write popular society, for without his presence 'it [society] would scarcely have had a vocabulary' (p. 133). Just as Vawdrey writes himself in two, Mellifont's 'writing' has transformative power so that social trends are reinvented. Mellifont not only rewrites social identity, shaping moods and styles in the fashionable world, he is also written himself. He is constantly compared to an actor who never forgets his lines, and whose part has been so carefully composed that it seems 'his very embarrassments had been rehearsed' (p. 196). Indeed, Mellifont's very being must be written by those around him, his lines foisted upon him, and without another's gaze he cannot contextualize himself. He is a 'legend', the perfect socialite, created through the public's ceaseless talk of his actions and sayings (p. 196). But in authoring the latest social trends, Mellifont, like Vawdrey with his creative writing, is at the mercy of the masses, and has nothing but a ghost of himself left for his private life.

James's story is particularly anxious about the possibility that in writing (whether Mellifont's symbolic writing of social fads or Vawdrey's fiction) the successful writer must literally become a stranger to himself in order to achieve critical and commercial prestige. In making this division, however, the writer is in danger of having his secret or private life discovered. In Mellifont's case, the terrible truth in his private life is that he does not have one (perhaps because there are so many windows in the house of fiction that the public can always see in), whereas Vawdrey's unsettling secret is that his public face is only an imitation of his private one. If Vawdrey's private life is discovered, then everyone will know that his public self has simply been mechanically, uncannily performing social duty. When Blanche Adney realizes the truth about the public Vawdrey, the narrator observes that 'she shrank from him, without a greeting; with a movement that I observed as almost one of estrangement' (p. 226). For her, by the end of the story, 'the other' Vawdrey (the private Vawdrey) becomes 'the real one' (p. 222). The price Vawdrey pays for artistic success is the sacrifice of the 'academic neatness' of his identity.

One of Freud's definitions of the uncanny is that of something which is familiar, or of the home, and also, paradoxically, unfamiliar and strange to the home (p. 220). Significantly, Mellifont 'would not be at home' with initiating a discussion about Vawdrey's private life, since it lies too closely with his own particular problem (p. 207). Furthermore, Lady Mellifont is 'not at home with him [Lord Mellifont]' herself (p. 413). The critic Paul Coates has suggested that one's lover is always one's other, or double – his 'intenser self', as James suggests (p. 213).[15] And yet Lady Mellifont's double has become completely unfamiliar. She both knows and refuses to register the truth about her husband, who is indeed not at home with himself or with her. Mellifont has to play the part of a devoted husband, implying that his identity is always a role carefully refined through artifice. He is threatening to the narrator, not because of his all-encompassing presence, but because of the absence that must lurk beneath: 'I [...] had wondered what blank face such a mask had to cover, what was left to him for the immitigable hours in which a man sits down with himself' (p. 213).

The revelations of Vawdrey's private life and Mellifont's public one, and the new familiarization with their identities in an unfamiliar place, indicate that the very choice of setting for the story is uncanny. The hotel is a transitional space for its guests and lacks any specific private or public areas: the narrator enters Vawdrey's chamber as easily as if it were his own. The hotel also evokes James's elaborate 'house of fiction' metaphor, which is filled with figures in every window looking out and each seeing something fresh or unexpected. While Lady Mellifont does not see (or perhaps refuses to see) that her husband has lost or forsaken his private life, the narrator surveys all aspects of him, both present and absent. In the context of this story James's 'house of fiction' metaphor becomes a metaphor not only for writing, but also for the mind. Thomas Hardy's discussion of the mind as a palimpsest in *Far From the Madding Crowd* (1874) is here useful. Hardy's narrator suggests that 'man, even to himself, is a palimpsest, having an ostensible writing, and another beneath the lines'.[16] The house of mind, as evidenced by

---

[15] Paul Coates, *The Double and the Other: Identity as Ideology in Post-Romantic Fiction* (New York: St. Martin's Press, 1988), pp. 10–11.

[16] Thomas Hardy, *Far From the Madding Crowd*, ed. by Robert C. Schweik (London: Norton, 1986), p. 189.

Vawdrey, can hold many guests or identities, and hides unexplored resources, inscriptions, and erasures: beneath the surface of Vawdrey's public self, his private self is still composing hidden meanings.

Multiple Personality, the Ghostly Mind, and James's Haunted House of Fiction

By the 1880s and 1890s, the understanding of the unconscious had radically changed from its perception in the mid-Victorian period, and writers like Henry James were negotiating with these new and often thrillingly unmappable theories of mind in their fiction. As Michael Davis has shown, the unconscious in the mid-nineteenth century was based on the 'model of the 'logical unconscious', which 'sees the conscious and unconscious minds as broadly alike in their activities, suggesting that the unconscious is, to some degree at least, knowable and predictable'.[17] Indeed, in the mid-Victorian period, the notion of knowing, predicting, and controlling aspects of the mind were the dominant modes of discussing the unconscious. In her discussion of Charlotte Brontë and contemporary psychology, for example, Sally Shuttleworth notes that

> the brash confidence of early Victorian optimism which gave rise to theories of moral management and the infinite malleability of the human psyche, gradually shades into the increasingly pessimistic visions of inherited brain disease and social degeneration propounded […] in the latter part of the century.[18]

It was not just fears about inherited disease and degeneration that marked understandings of the late-nineteenth-century unconscious, however, but also fears that the mind could no longer be managed. The earlier theories that the mind was neatly doubled or split between the conscious and unconscious had given way to the notion that the mind had *many* selves within.[19] Ian Hacking argues

---

[17]   Michael Davis, *George Eliot and Nineteenth-Century Psychology: Exploring the Unmapped Country* (Aldershot: Ashgate, 2006), p. 147.

[18]   Sally Shuttleworth, *Charlotte Brontë and Victorian Psychology* (Cambridge: Cambridge University Press, 1996), p. 37.

[19]   For more on double consciousness in the early and mid-Victorian period, see, for example, *Embodied Selves: An Anthology of Psychological Texts, 1830-1890*, ed. by Jenny Bourne Taylor and Sally Shuttleworth (Oxford: Clarendon Press, 1998), and Anne Harrington, *Medicine, Mind and the Double Brain: A Study in Nineteenth-Century Thought* (Princeton: Princeton University Press, 1987). In 1816, Mary Reynolds 'became one of the most famous instances of double consciousness during the nineteenth century' (Robert Macnish, 'The Case of Mary Reynolds' [1830], in Taylor and Shuttleworth, eds, *Embodied Selves*, pp. 123–4 [p. 123]). In 1856, Felida X. was brought to the public's attention as another sufferer of split selfhood. For more on multiple personality at the end of the century, see, for example, Hippolyte Bourru and P. Burot, *Variations de la Personnalité* (Paris: Baillière, 1888); Bourru and Burot, 'Un cas de la Multiplicité des états de Conscience Chez

that multiple personality 'came into being' in 1885 when the patient Louis Vivet was found to possess eight separate personalities.[20] Indeed, during the 1880s and 1890s, many scientists began to address multiple personalities in their writings on the mind. Jill Matus discusses Stevenson's *Jekyll and Hyde* 'in relation to research on conscious and unconscious or automatic processes and the concept of "multiplex personality", particularly as it was articulated in the 1880s by Myers'.[21] In doing so, she 'endeavor[s] to articulate and respond to the crisis of subjectivity and identity raised by the idea of the multiple self'.[22] While I am also interested in changing perceptions of identity, I want to highlight mental scientists' impossible desire to structure the mind with defined limits and clear conceptions of self during a time when they were also assigning the mind limitless possibilities. Although Hacking suggests that late-Victorian mental scientists were 'attempt[ing] to scientize the soul through the study of memory', he ignores the inconsistencies of systematizing identity when notions of selfhood were becoming increasingly inchoate by the century's close.[23]

The example of nineteenth-century physicians Hippolyte Bourru and Prosper Burot demonstrates the ways in which mental scientists were attempting to rigorously control the science of mind. Bourru and Burot believed that multiple personalities could be grounded within memory: '[t]he comparison of previous states of consciousness with present states is the relation that unites a formed psychic life with the present one. That is the foundation of personality. A consciousness that compares itself with a former one is a true personality'.[24] Bourru and Burot's use of the term 'foundation' is significant here as, like William James, they evoke the mind's architecture. While Bourru and Burot's language suggests their belief that the mind and identity can be comfortably housed within an academic and impermeable structure of selfhood, William James sees the problem with this

---

un Hystéro-Epileptique', *Revue Philosophique*, 20 (1885), 411–16; John H. King, *Man an Organic Community: Being an Exposition of the Law that the Human Personality in all its Phases in Evolution, Both Co-ordinate and Disconsolate, is the Multiple of Many Sub-Personalities*, 2 vols (London, 1893); A.T. Myers, 'The Life-History of a Case of Double or Multiple Personality', reprinted from *The Journal of Mental Science* (1886); F.W.H. Myers, 'Multiplex Personality' (1886), in Taylor and Shuttleworth, eds, *Embodied Selves*, pp. 132–8.

[20]   Ian Hacking, *Rewriting the Soul: Multiple Personality and the Sciences of Memory* (Princeton: Princeton University Press, 1995), pp. 180–83 (p. 180).

[21]   Jill Matus, *Shock, Memory and the Unconscious in Victorian Fiction* (Cambridge: Cambridge University Press, 2009), p. 19.

[22]   Ibid., p. 182.

[23]   Hacking, p. 6.

[24]   Quoted in Hacking, p. 180, from Bourru and Burot, *Variations de la Personnalité*. Bourru and Burot use different states of consciousness as a means of talking about the manifestations of different personalities in a case of multiple personality disorder. The physicians coined the term 'multiple personality' in 1885 when they were studying Louis Vivet.

theory: by the 1880s and 1890s, the architecture of the mind was overrun, its structure incoherent and ambiguous; 'few outlines are pure'.[25]

At the end of the century, mental scientists were anxiously policing boundaries of the self and mind in unprecedented ways, despite the fact that cases of multiple personality only drew attention to the impossibility of mapping and delimiting identity. Indeed, the connections made between the workings of the mind and the supernatural by mental scientists and SPR members indicate not only that mental scientists strove for an unobtainable level of academic specificity within the studies on the mind, but also that the foundations of identity had become incorporeal and ghostly. Their activities serve as examples of the ways in which studies on the supernatural and psychology worked in tandem in the late nineteenth century. For example, while never taking up a profession, Edmund Gurney studied both psychology and medicine and published for the SPR. In *Phantasms of the Living* (1886), which he co-wrote with F.W.H. Myers and Frank Podmore, Gurney suggested that a telepathic message from the mind could manifest itself in the form of a ghostly apparition.[26] William James was interested in the effect of ghosts, the supernatural, and phenomena like telepathy on the human mind. William McDougall was a psychologist interested in instinctual and social behaviour and eugenics. He was also interested in parapsychology, and became the president of the SPR in 1920–1921 and the American SPR in 1921. French physiologist Charles Richet won the Nobel Prize in 1913 for his work on anaphylaxis. President of the SPR in 1905, Richet investigated paranormal activity and coined the term ectoplasm (the white substance exuded by mediums during a séance, which purportedly materializes into spirits).

Scientists and SPR members saw particularly strong connections between multiple personality and ghosts. As Hacking points out, psychic research on spiritualism led to the hypothesis that 'alters [alternate personalities in cases of multiple personality] were departed spirits; mediumship and multiplicity grew close'.[27] Myers's *Human Personality and Its Survival of Bodily Death* (1903) examines this very phenomenon, suggesting that alternative personalities are in fact the manifestations of ghosts attempting to contact the living.[28] The mind could host both other selves and ghosts, defying logical reason. Mental scientists are haunted by the immaterial nature of mind, their attempts to produce a 'materialist science of the self', as Jenny Bourne Taylor and Sally Shuttleworth phrase it in *Embodied Selves*, instead 'express[ing] a profoundly ambivalent sense of self'.[29]

[25]   William James, 'Frederic Myers's Service', p. 156.

[26]   Gurney, Edmund, F.W.H. Myers, and Frank Podmore, *Phantasms of the Living*, 2 vols (London: Trübner, 1886).

[27]   Hacking, pp. 135–6.

[28]   F.W.H. Myers, *Human Personality and Its Survival of Bodily Death*, 2 vols (London: Longmans, 1903).

[29]   Taylor and Shuttleworth, eds, p. xiii; p. xiv.

While certainly 'academic neatness' in William James's 'world of mind' was lost, the field had gained 'romantic improvers' who opened up new lines of investigation into the mind.[30] Henry James's 'The Private Life' is influenced by fin-de-siècle theories of the mind, particularly the haunting possibilities of multiple personality, but the story also suggests anxiety about where the boundaries of identity lie in another way: Who owns the rights to the public and private selves, and to the writing itself? Is it the author, another self within, another separate, ghostly self, or the reading public that ultimately has agency in the act of writing?

In 'The Private Life' James tries to circumvent these fears by attempting to make sharp distinctions between public and private identities. While this division uncannily doubles the self, it also artificially seems to offer James a means of imposing 'academic neatness' on his text: there is one public Vawdrey and one private Vawdrey. This attempt quickly disintegrates, however, when it seems clear that there are already other selves (or personalities) within the text. 'The Private Life' is haunted by a ghostly third presence whom the other characters think they know, a hybridized mixture of the public and private selves. Vawdrey's identity is dispersed throughout the hotel: like James's 'house of fiction' metaphor in which guests stand at every window, Vawdrey's selves manifest themselves in the various chambers, writing in darkened rooms and socializing in salons. The hotel is comparable to the mind itself, populated by ghostly other selves but lacking solid structure, with obscure corridors (p. 204) and 'vague [...] apertures', with 'flitting illumination' (p. 205).[31] It is the apparitional nature of identity that haunts and fascinates James, who yearns to restrain the self within concise and unadulterated prose, yet realizes the ultimate futility of such a project.

'The Private Life' expresses many of the anxieties James himself felt about the effect of writing on his own private circumstances. In 1909 he burned much of his correspondence, destroying the discursive records of a lifetime.[32] Lyndall Gordon suggests that this was an attempt to resist biography, and therefore adverse judgement by the public (both his contemporaries and future readers).[33] Certainly it seems that James hopes to bury biography so that his private life cannot be

---

[30]   William James, 'Frederic Myers's Service', p. 156.

[31]   The 'flitting illumination' may refer to early kinds of moving pictures like the praxinoscope (c. 1877) which, according to the OED, is '[a] scientific toy resembling the zoetrope [c. 1867], in which a series of pictures, representing consecutive positions of a moving body, are arranged along the inner circumference of a cylindrical or polygonal box open at the top, and having in the middle a corresponding series of mirrors in which the pictures are reflected; when the box is rapidly revolved, the successive reflexions blend and produce the impression of an actually moving object'. James may be using the multiple, unstable images in early cinematographic technology as symbols for the multiple and unstable representations of identity in 'The Private Life'.

[32]   Fred Kaplan, *Henry James: The Imagination of Genius: A Biography* (London: Hodder & Stoughton, 1992), p. 521.

[33]   Lyndall Gordon, *A Private Life of Henry James: Two Women and His Art* (London: Norton, 1999). Gordon suggests that James was not only inspired by women in his life, but

claimed or misinterpreted by voyeuristic readers and unscrupulous publishers. However, in killing off biography, James might also be resurrecting literary fame, since his destruction could incite public interest. It does appear that James is anxious about who inherits the haunted house of fiction: When James dies, who will possess the deed to his writing and unlock its secrets? How can he maintain 'academic neatness' in his work once it has passed into the public sphere, beyond his grasp? 'The Private Life' reveals that the literary marketplace has an uncanny effect on writers. The ghostly third in the story might also be the spectre of celebrity that haunts the author in its promise of literary greatness and fame and its potential poisoning of the author's private life.[34] For James, writing is a cycle of the conjuring and exorcism of ghosts: the unsettling revelation of his private life, and then the enshrouding of his personage again.

The High Priestess of the Remington: James, the Typewriter, and Agency

> Forgive a communication very shabby and superficial. It has come to this that I can address you only through an embroidered veil of sound. The sound is that of the admirable and expensive machine that I have just purchased for the purpose of binding our silences. The hand that works it, however, is not the lame *patte* which, after inflicting on you for years its aberrations, I have now relegated to the shelf, or at least to the hospital – that is to permanent, bandaged, baffled, rheumatic, incompetent obscurity.[35]

James's letter to W. Morton Fullerton on 25 February 1897 was among the first that he ever dictated, after writer's cramp in 1896 forced him to hire an amanuensis and buy a typewriter. While the letter seems to endorse the 'admirable' typewriter, it also reflects on how using a typewriter changes the nature of communication. In this case, communicating from a distance with an old friend means James's writing is not only in danger of becoming 'shabby and artificial' with the intervention of a typewriter, but it is also no longer a means of directly engaging with Fullerton. For James this mediated process is haunted, the letter's language and imagery implying supernatural communication over distance and the spiritualist séance.[36]

---

also that he 'fed' on them to sustain his art (p. 304). In Chapter 5, I suggest a different kind of relationship between James and Vernon Lee, who were haunting each other's writing.

[34]   See also Collin Meissner, '"What ghosts will be left to walk": Mercantile Culture and the Language of Art', *Henry James Review*, 21 (2000), 242–52. Meissner suggests that James's 'aesthetic sensibilities' were horrified by 'the essential hollowness or vacancy behind the façade of American mercantile and social culture' (p. 244). Although Meissner focuses on mercantile culture, his article is significant to an understanding of James's anxieties about the literary marketplace.

[35]   Henry James, 'To W. Morton Fullerton', in *Henry James Letters*, ed. by Leon Edel, 4 vols (London: Macmillan, 1974–1984), IV (1984), pp. 41–2 (p. 41).

[36]   Mark Seltzer's 'The Postal Unconscious', *Henry James Review*, 21 (2000), 197–206, problematically argues that James was comfortable with using new technology.

The notion that the typewriter can 'bind their silences' implies a kind of spiritual or ecclesiastical binding, the magical power of a spell, or even a telepathic link in which James is uncannily connected to Fullerton over a great distance. The word also implies that James wishes to symbolically cover the wounds made by their periods of non-communication with the aid of the typewriter. While the typewriter seems to promise intimacy over distance, James is anxious about the fact that he may be even further from attaining direct communication, his words separated from Fullerton not only spatially, but also by the bodies of the typewriter and the typist.

Indeed, the 'veil of sound' through which James now communicates suggests that the sound of the typewriter prevents him or distracts him from really communicating with Fullerton, and also links him with the ghosts of spiritualist séances. The 'veil of sound' could be associated with the audible manifestations during a séance in which trumpets are played by spirits above the sitters' heads, wreathing them in a cacophony of sound. The 'veil' also seems suggestive of the symbolic veil spiritualists use to discuss various aspects of spiritualism, such as the separation between this world and the next.[37] When James dictates, he becomes the spirit channelling his thoughts through the medium (the typist/typewriter), which then filters through to the sitters at the séance (Fullerton).

James tells Fullerton that 'the hand' working the typewriter is not his own, but another's. While James means to convey that he has hired a typist, the notion of an amanuensis makes issues of authorship problematic. Who can be credited as the author?[38] That James has 'relegated' his hand to 'obscurity' is significant here because it suggests that he finds the act of writing by hand obscure and anachronistic, deficient in both clarity and intelligibility but also lacking currency: writing by hand is technologically out of date, 'bandaged, baffled, rheumatic, [and] incompetent'.

Here James turns away from the private writing he describes in 'The Private Life' and towards a more public style of composition, in which amanuensis and typewriter are witnesses to the creative process. As 'The Private Life' shows, writing by hand is a ghostly private act, but dictating to a typist is haunting in a new way. The typist and the typewriter become part of the writing ritual, the mediums to James's message, and what filters through might not always be entirely

---

I suggest that James was uneasy about switching to the typewriter, and that this new process of writing was an uncanny one.

[37] For more on the use of the term 'veil' in spiritualism, see, for example, John Traill Taylor, *The Veil Lifted: Modern Developments in Spirit Photography* (London: Whittaker & Co., 1894), originally published in 1894; and W.T. Stead, *The Blue Island: Experiences of a New Arrival Beyond the Veil* (London: Rider, c. 1922). Stead's book is suggestive about the late Victorian preoccupation with the ghostly nature of writing and questions of agency – he dictated the work after his death to two mediums, who transcribed and published his experiences of the world 'beyond the veil'. Nathaniel Hawthorne's *The Blithedale Romance*, first published in 1852, describes a 'veiled lady', a mysterious, ghostly woman said to possess mesmeric power.

[38] See Katherine Rowe, *Dead Hands: Fictions of Agency, Renaissance to Modern* (Stanford: Stanford University Press, 1999), for a discussion of agency and hands.

James's creation. Not only might the typist alter his words, but her very presence might surreptitiously influence him.

The uncanny nature of writing technologies suggests that James's hopes that the typewriter can restore academic neatness to his writing and exorcise his ghosts are futile. He insists that using the typewriter gives him greater editorial control over his work, but that otherwise the process is the same as if he were writing by hand:

> [t]he value of that process for me is in its help to do over and over, for which it is extremely adapted, and which is the only way I can do at all. It soon enough becomes *intellectually*, absolutely identical with the act of writing – or has become so, after five years now, with me; so that the difference is only material and illusory.[39]

Yet, while James remarked on the efficacy of using a typewriter, Leon Edel observes that there are many notable differences in his style once he begins dictating:

> In this process certain mannerisms crept in – attempts to get away from old familiar forms of expression: displacement and splitting of verbs, the emergence of unexpected adverbs, the removal of the given phrase from that part of the sentence where the reader expected to find it, into another part. James's prose was now spoken prose.[40]

James's style seems to *lose* its academic neatness, and the language and syntax become almost haphazard. Still, the writing is not just 'spoken prose', as Edel suggests, but rather a complex amalgam of the permanency of print and the spontaneity of speech, a blurring of the distinctions between written and oral communication. The typewriter gives James even less control over his 'house of fiction' (and its already shaky foundations) and further destabilizes the notion of authorship and agency.

'[C]ertain mannerisms creep in' to James's writing at this period, but crucially for James it is the possibility for other *personalities*, those of his amanuenses, to creep into his work which preoccupies him. James had three primary amanuenses: William MacAlpine, Mary Weld, and Theodora Bosanquet (who worked for James from 1907 until his death in 1916). According to James, MacAlpine 'had too much Personality – and I have secured in his place a young lady [Weld] who has, to the best of my belief, less, or who disguises it more'.[41] MacAlpine's overbearing personality contrasted with Weld's meeker one is suggestive about conventional expectations for gender roles, both in the workplace and in society at large in the nineteenth century, but for James the real problem was the possibility that his employees might have too much sway over his writing: 'Mr. James liked his

---

[39]   Leon Edel, *Henry James: The Master 1901–1916* (London: Hart-Davis, 1972), p. 130.
[40]   Ibid., p. 130.
[41]   Ibid., p. 94.

typists to be "without a mind" – and certainly not to suggest words to him'.[42] James's emphasis on MacAlpine's 'Personality' implies not only that MacAlpine has a strong character, but that he has an *overly* strong character, which might drown out James's own sense of creative potential.

James described Bosanquet as

> a new excellent amanuensis, a young boyish Miss Bosanquet, who is worth all the other (females) that I have had put together and who confirms me in the perception afresh – after eight months without such an agent – that for certain, for most kinds of diligence and production, the intervention of the agent is, to *my* perverse constitution, an intense aid and a true economy![43]

Bosanquet occupies the uneasy position between automatic writer, mindlessly copying James's words, and active agent in the writing process. Significantly, Lisa Gitelman has shown that 'during the 1890s "automatic writing" was a phrase that applied doubly to the work done on typewriters and during séances, by secretaries and mediums, both of whom were usually women'.[44] Indeed, Bosanquet filled both positions: after James's death she became an automatic writer and medium and even corresponded with James's spirit. As James's typist, Bosanquet channels his words through the typewriter so that his story can manifest itself on the page. Superficially, it seems as if Bosanquet is passively, automatically recording James's words. Indeed, despite cultural shifts in attitudes towards the acceptability of women in the workplace, women's work was often repetitive, stifling creative and intellectual endeavour.[45] Many women typists, telephone operators, and telegraphers acted to transmit other people's messages, and never to author their own.

---

[42]   Ibid., p. 367.

[43]   Ibid., p. 370.

[44]   Gitelman, p. 19.

[45]   The increase in the number of working women during the second half of the nineteenth century was due to two factors. Firstly, women could be paid less than men for doing the same job (Gitelman, p. 188). Henry James, for example, was pleased to hire Mary Weld once William MacAlpine left his employ because MacAlpine 'is too damned expensive, and always has been […] I can get a highly competent little woman for half' (Edel, *Master*, p. 91). Secondly, the new technologies seemed suited to feminine capability according to gender constructs of the period. Tom Standage's description of female telegraphers in Britain is useful in the context of social expectations about 'women's work' during this period, as well as an interesting summary of the 'average' female telegrapher: 'female telegraphers were usually the daughters of clergymen, tradesmen and government clerks, and were typically between 18 and 30 years old and unmarried. Women were regarded as "admirable manipulators of instruments", well suited to telegraphy (since it wasn't too strenuous), and they could spend the quiet periods reading or knitting. The hours were long, though; most operators, including the women, worked ten hours a day, six days a week'. From Standage, *The Victorian Internet: The Remarkable Story of the Telegraph and the Nineteenth Century's Online Pioneers* (London: Weidenfeld and Nicolson, 1998), pp. 125–7. The typewriter reached the public in 1874, and when, by the 1880s and 1890s,

The relationship between Bosanquet and James, however, challenges this dynamic. In *Literature, Technology and Magical Thinking*, Pamela Thurschwell discusses James's 'In the Cage' and Bosanquet, suggesting that together they 'explor[e] potential pleasures and dangers of carrying the words of others'.[46] While Thurschwell is particularly concerned with sexuality, and the economics of exchange in James, she does not go far enough to examine the complex ways in which James approached the act of writing, how the typewriter changed and made uncanny the ways in which he wrote, and how Bosanquet herself made problematic the dynamic of the passive secretary being dictated to by the male employer. Although Bosanquet is described as an 'intense aid and a true economy', her intensity implies she is not only very helpful, but violently, even excessively so.[47] She is also part of the economy of supply and demand, linked to the marketplace, which disturbs James since the marketplace (like the typist herself) might control, influence, corrupt, or reject his writing. James's description of her as an agent grants her power in the writing process, and she symbolically becomes his literary agent, negotiating between the public and the private. Paradoxically, he also refers to her presence as an 'intervention', or an interference.[48] Bosanquet both helps and hinders James's writing process, possibly because James fears that to a certain extent the words on the page, filtering through her, are somehow altered, the very writing itself modified by her personality.

From the male-only Reform Club in London, James wrote to Bosanquet on 27 October 1911 that 'I haven't a seat and temple for the Remington and its priestess'.[49] The term 'priestess' implies that Bosanquet possesses spiritual and mystical capabilities in her role as an automatic writer. Perhaps James *wants* Bosanquet to become a priestess of celebrity, protecting the 'temple' (or house) of fiction from the tainted literary marketplace. Although James wants Bosanquet to uphold 'academic neatness', nevertheless her presence during scenes of writing problematizes his desire for unmediated authorship, and he is left feeling anxious that literary agency is in flux between himself, Bosanquet, and the typewriter.

Uncanny Technologies of Writing at the Fin de Siècle: Kipling's 'Wireless' and James's 'In the Cage'

Discursive technologies at the end of the century raised troubling questions about literary agency and influence (as in the case of James and his typewriter), but they also brought up concerns about the effects of interacting with these

---

the typewriter was in common use in businesses everywhere, the female secretary (the woman automatic writer), was employed en masse (Gitelman, p. 188).

[46]   Pamela Thurschwell, *Literature, Technology and Magical Thinking, 1880–1920* (Cambridge: Cambridge University Press, 2001), p. 87.

[47]   Edel, *Master*, p. 370.

[48]   Ibid., p. 370.

[49]   Henry James, 'To Theodora Bosanquet', in *Henry James Letters*, IV, p. 589.

writing machines. What impact did using technologies of writing have on the understanding and construction of identity? By examining contemporary theories about telegraphy and telepathy and offering close readings of Rudyard Kipling's 'Wireless' and James's 'In the Cage', I suggest that bodies, minds, and machines at the end of the century had the uncanny ability to mingle together. Minds worked like machines, machines evoked and retracted physicality and bodily proximity, bodies touched machines to make them work, and machines tapped the workings of the mind, acting as the mediums for discursive practice. Minds could touch other minds telepathically, just as machines could touch other machines to transmit messages electrically. Spiritualist mediums acted as mind, body, and machine combined, utilizing new technology like the Morse code to transmit messages and becoming automatic writing machines like the typewriter to bodily transmit writing, whether from spirit minds or their own unconscious minds.

Terry Castle has already linked bodies and machines to the uncanny, pointing to the eighteenth-century interest in technology, particularly in the form of automata, in order to argue that the 'eighteenth-century invention of the automaton was also an "invention" of the uncanny'.[50] But the uncanniness of interactions with machines is of a different nature by the time we reach the fin de siècle, whose discursive cultural practice renders this change. The merging of bodies and machines at the turn of the century, their interchangeability, also includes the mind, in part because of contemporary notions about mental science, psychical research, and discursive technologies. This blending of discourses suggests that identity at the fin de siècle could come into dangerous proximity with other identities: the self might be vulnerable to outside influence (a theme I will return to in my discussion of mesmerism in Chapter 3).

Perceptions about the wireless telegraph (c. 1894–1896) and telepathy, for example, illustrate the ways in which machines, bodies, and minds were conceptually linked. If electrical impulses could be sent through the air to transmit telegrams, scientists theorized that telepathy was also possible – thought waves could be sent out and received by distant bodies and minds. In 1882 Myers introduced the term 'telepathy', defining it as 'cases of impression received at a distance without the normal operation of recognized sense organs'. Indeed, if telepathy and telegraphy involved receiving an 'impression' from a distance, how did this outside influence affect identity? Myers's definition of telepathy suggested that the wireless telegraph and thought transference could both be classified in the same category.[51] Minds and machines were beginning to echo one another's

---

[50] Terry Castle, *The Female Thermometer: Eighteenth Century Culture and the Invention of the Uncanny* (Oxford: Oxford University Press, 1995), p. 11.

[51] Myers et al., 'First Report of the Literary Committee', in *Proceedings of the Society for Psychical Research*, vol. I, pt. 2 (London: Trübner, 1883), p. 147, cited in Nicholas Royle, *Telepathy and Literature: Essays on the Reading Mind* (Oxford: Blackwell, 1990), p. 2. Luckhurst's *The Invention of Telepathy* discusses telepathy in detail as it was received in the 1880s and 1890s.

workings, both invested with limitless and haunting possibilities, and conceptually developing co-dependently in the eyes of the public as a result of the interchange of meaning they provided for one another.

Laura Otis has shown that nineteenth-century technologies were necessarily influenced by popular culture, and also used the terminology of human anatomy.[52] The telegraph network, for example, was talked and written about in the same way that one would discuss the nerve impulses of the brain, with electrical impulses relaying messages from (nerve) centre to (nerve) centre. If the language of the body was to affect the language of technology, then bodies and technology were, from the beginning, very much interconnected. Myers's telepathy seems to indicate such a relation, since the etymology of the word itself implies a connection between telepathic machines and the human body. Telepathy comes from the Greek *tele-*, meaning distance, and *-pathy*, meaning feeling or perception. However, Luckhurst also shows that *-pathy* comes from *pathos*, connoting both intimacy and touch.[53] This slight change in definition allows telepathy to mean bodily intimacy over distance, a communion of minds that is also a physically psychic embrace. The physicality within the term can also be linked to 'telepathic' machines like the wireless. Humans work the machines, the oil from their fingers greasing handsets and switchboards, hair and skin follicles left behind as a part of the process of connecting flesh and mechanical design. Mechanical automatic writers like typewriters were marked by the body, whereas human automatic writers like mediums left bodily secretions such as ectoplasm during séances.

And yet the machine, despite its intense closeness to the body, is irrevocably separate from its human counterpart.[54] Although James hoped that communicating with Fullerton through the typewriter would bind them more closely together, the typewriter only seemed to make their communication even more superficial and distant. In Kipling's 'Wireless', the interaction between mind, body, and the telegraph machine also disrupts and dislocates, rather than enhances, communication.

Significantly, 'Wireless' is filled with discursive figures: wireless telegraph machines, human automatic writers, and the Romantic poet Keats.[55] These discursive machines, bodies, and minds are conflated within the story, shattering identity and blurring thresholds of self and non-self. Indeed, who is body,

---

[52]   Laura Otis, *Networking: Communicating with Bodies and Machines in the Nineteenth Century* (Ann Arbor: University of Michigan Press, 2001), p. 2.

[53]   Luckhurst, *The Invention of Telepathy*, p. 1.

[54]   See Connor, *Dumbstruck*, for a discussion of the ways in which technology like the telephone, which gave the listener the impression that the speaker was almost in the ear, suggested an uncanny intimacy between bodies that were actually distant (pp. 364–8).

[55]   Rudyard Kipling, 'Wireless', in *The Best Short Stories* (Ware: Wordsworth Classics, 1997), pp. 143–58. Subsequent references to this edition are given after quotations in the text. The story refers to the Marconi experiments in wireless telegraphy of the fin de siècle. Gugliemo Marconi was the Italian inventor and developer of the wireless telegraph.

machine, or mind is a complicated problem in the story. Bodies in 'Wireless' can be confused with other bodies: Fanny Brand might actually be Fanny Brawne, and Shaynor the druggist becomes the medium for Keats the druggist.[56] Machines can be confused with other machines; for example, the messages from the 'men o' war' ships are picked up by the Morse instrument in the druggist's, but not by each other (p. 157). Finally, minds can be confused with other minds: the narrator's unconscious mind draws parallels between the evening's events and the lines of Keats's 'Eve of St. Agnes' (1819), and then transfers these connections, seemingly telepathically, into Shaynor's mind. However, Shaynor himself seems to be the combination of mind, body, and machine. Shaynor is an automatic writer – unconsciously, mechanically writing the words of the dead poet Keats, his fleshly body confused with a mechanical one. Shaynor's mind and the disembodied mind of Keats also become confused, intermingling with each other – the mind of the dead poet entering into the mind of Shaynor's 'machine-like' body (p. 152).[57] And certainly Shaynor's body is bodily, secreting bodily fluid in the form of the 'bright-red danger signals' (p. 147) – the blood of the consumptive. Shaynor here seems telepathically, empathetically ill with Keats. But if Shaynor is acting as mind, body, and machine for the outpourings of Keats's poetry, then he also seems to be acting in parallel to the signals passing from wireless to wireless – he himself is a wireless telegraph machine. The narrator remarks, 'there is something coming through here, too' (p. 152), suggesting that Shaynor himself has been transformed into a wireless, his 'start[s]', 'wrench[es]', and 'jerk[ing]' movements like the tickings and stoppings of the Morse code (p. 148).

The meeting point between minds, bodies, and machines in 'Wireless' is the uncanny moment when identity fragments.[58] Shaynor blurs into both the wireless and the dead poet, and as the narrator watches Shaynor's transformation into that which is not entirely human, he finds himself splitting apart, his own identity merging with that of the technology around him: 'I heard the crackle of the sparks as he [Cashell] depressed the keys of the transmitter. In my own brain too, something crackled' (p. 152). At this point, when the narrator can no longer distinguish between mind, body, and machine, he becomes doubled: 'For an instant that was half an eternity, the shop spun before me in a rainbow tinted whirl, in and through which my own soul most dispassionately considered my own soul that fought with an over-mastering fear' (p. 152). The narrator dissociates

---

[56] Keats was actually a druggist as well as a poet. His life is echoed in other ways in 'Wireless': he became engaged to Fanny Brawne in 1819, the same year that he wrote 'The Eve of St. Agnes', and died of tuberculosis in 1821.

[57] Kipling may also be playing with the notion of the afflatus, which imparts poetic inspiration: Keats is the muse who is mediated by the wireless.

[58] Luckhurst argues that because Kipling uses technologies, 'the most visible products of modernity', in his occult stories, the uncanny is an inappropriate analytical tool (*The Invention of Telepathy*, p. 177). I suggest, however, that the uncanny is crucial to understanding the ways in which the mingling of minds, bodies, and machines alters conceptions of identity in the 1880s and 1890s.

himself so completely from his 'other self' (p. 152) that he can actually speak to it, suggesting the existence of a doubly uncanny moment: not only has the self uncannily separated from itself, but also, one of the selves perceives the uncanny truth that there may be no scientific, rational explanation for Shaynor's 'possession'. Because selfhood in this story has no distinctive barriers between what is and what is not one's own identity, identity is free to spill over from its unified form into splinter selves.

In a text about bodies, minds, and machines merging together, invading one another and disregarding normal boundaries of physical separation, Kipling's use of Keats's 'The Eve of St. Agnes' (1820) becomes particularly significant: Keats's poem is the story of Porphyro's rape of Madeline, of bodies penetrating other bodies, just as the poem itself penetrates the text of 'Wireless'.[59] In this context, telepathy becomes invested with a degree of perverse sexuality. For example, Shaynor's 'start[s]', 'wrench[es]', and 'jerk[ing]' (p. 148) suggest ugly and unsettling sexual behaviour. 'Wireless' also touches on fin de siècle anxieties about sexuality: what does it mean when male bodies can be possessed by other male bodies?

Significantly, telepathy in 'Wireless' becomes possessed of a haunting ability in which the projection of thoughts into the mind of another suggests not merely a rape of the mind, but the power of discursivity itself. Telepathy is, after all, a discursive practice, the reading of transmitted thoughts and the writing and transmission of them for others to read. Keats transmits his thoughts to Shaynor, haunting him, and his 'The Eve of St. Agnes' haunts the short story itself, leaving spectral traces of the poem's 'hare', 'moths', and the colours red, yellow, and black throughout 'Wireless'. The writer has the power to resurrect himself, to infiltrate Shaynor's mind, but also the minds of future readers of both Kipling's 'Wireless' and Keats's works. Keats's poem 'This Living Hand' (1819) seems a fitting prophecy for Keats's resurrection in Shaynor's body.[60] The poem suggests that Keats's hand 'would, if it were cold' and dead, 'haunt thy days and chill thy dreaming nights', which it successfully does in 'Wireless'.[61] Cashell's conjuring of the image of a 'spiritualistic séance' in conjunction with the night's jumbled, mis-wired reception of messages is apt indeed (p. 158). 'Wireless' becomes an evocation of the power of writing itself, which, ghost-like, haunts readers. Writing in 'Wireless' is manifested with uncanny

---

[59]   John Keats, 'The Eve of St. Agnes', in *Complete Poems: John Keats*, ed. by Jack Stillinger (London: Belknap Press, 1982), pp. 229–39.

[60]   Keats, 'This Living Hand', in ibid., p. 384. In *Dead Hands*, Rowe suggests that Keats's dead hand symbolizes the 'Romantic alienation of the subject from the work of his or her hands' (p. 114). Spiritualist automatic writings of the Victorian period, however, transform the Romantic uneasiness about accepting their writing as their own into the Victorian conviction that what has been written is always at risk of being the work of someone else.

[61]   Keats, 'This Living Hand,' in *Complete Poems*, p. 384.

capability, fragmenting identities in blurred relations of bodies, minds, and machines and immortalizing itself by returning from the dead again and again.[62]

The séance-like communion with a dead poet and distant ships in 'Wireless' gives the reader 'just enough to tantalize', a hint at what might be possible in this collision of writing, discursive technology, the human mind, and the human body (p. 158).[63] James's 'In the Cage', another story about telegraphy, is also fascinated by that which can 'tantalize'. The 'odds and ends' Cashell finds so 'disheartening' (p. 158) are the pieces of information upon which the girl in 'In the Cage' thrives. She longs for glimpses into the lives of high society, for information that can be 'patched up and eked out' into a narrative that will signify for her the 'high reality' and 'bristling truth' of the wealthy who regularly send and receive telegrams.[64] She desires 'a play of mind' (p. 144) that allows her to 'piece together all sorts of mysteries' (p. 145), the mysteries haunting the lives of the gentlemen and ladies that come to her, a kind of 'knowing' that 'made up for the long stiffness of sitting there in the stocks' (p. 143), and, in effect, makes up for all that is unsatisfactory and dull in her life. She develops 'flickers of antipathy and sympathy' (p. 142), suggesting her development of telepathy in her 'surrender [of] herself [...] to a certain expansion of her consciousness' (p. 143), through which she gleans 'red gleams in the grey, fitful awakings and followings, odd caprices of curiosity' (p. 142). Like the telegraph machine which receives and transmits messages to people hundreds of miles away, the girl has 'an extraordinary way of keeping clues' (p. 145), receiving knowledge as it 'floats to her through the bars of the cage' (p. 146) and apprehending unspoken truths from a distance. Being a telegraphist allows her, as she puts it, to 'read into the immensity of their intercourse stories and meanings without end' (p. 155).[65]

---

[62]   See Sylvia Pamboukian, 'Science, Magic and Fraud in the Short Stories of Rudyard Kipling', *ELT*, 47 (2004), 429–45. Although Pamboukian is interested in the 'apparent opposition' between the supernatural and technology, showing how Kipling returns again and again to these themes, she overlooks the uncanny meeting of bodies and machines in his work (p. 429).

[63]   Kipling was fascinated by the ghostly and the supernatural, and often returned to these themes in his writing. He was familiar with the SPR, and his sister, Trix Kipling, was a renowned clairvoyant. For more on Kipling's connections to spiritualism, see, for example, William B. Dillingham, 'Kipling: Spiritualism, Bereavement, Self-Revelation, and "They"', *ELT*, 45 (2002), 402–25. For a critical study on Kipling and the supernatural, see, for example, Peter Morey, 'Gothic and Supernatural: Allegories at Work and Play in Kipling's Indian Fiction', in *The Victorian Gothic: Literary and Cultural Manifestations in the Nineteenth Century*, ed. by Ruth Robbins and Julian Wolfreys (Basingstoke: Palgrave, 2000), pp. 201–17.

[64]   Henry James, 'In the Cage', in *The Complete Tales of Henry James*, ed. by Leon Edel, 10 vols (London: Hart–Davis, 1962–1964), X, pp. 139–242 (p. 146). Subsequent references to this edition are given after quotations in the text.

[65]   For more on the girl's ability to discover hidden truths from her customers, see, for example, Eric Savoy, '"In the Cage" and the Queer Effects of Gay History', *NOVEL*, 28

In effect, the girl believes herself to have achieved a kind of omniscience about the lives of her 'clients', a telepathic knowing through telegraphy, itself a kind of telepathic machine, and this knowledge gives her the feeling of omnipotence:

> she had at moments, in private, a triumphant, vicious feeling of mastery and power, a sense of having their [the ladies and gentlemen of telegrams] silly, guilty secrets in her pocket, her small retentive brain, and thereby knowing so much more about them than they suspected or would care to think. There were those she would have liked to betray, to trip up, to bring down with words altered and fatal. (p. 154)

And yet as she begins to live 'more and more in the whiffs and glimpses', and finds 'her divinations work faster and stretch further' (p. 152), the power she believes she has gained is only an illusion. Whereas the wireless telegraph and telepathy in 'Wireless' offer infinite and terrible possibilities for the capabilities of the human mind, even suggesting that the writing mind can transcend death, telegraphy and telepathy in 'In the Cage' entangle the anonymous girl in telegraph wires of powerlessness. The metal lines of the cage encasing the girl symbolize the metal lines of the telegraph network that trap her socially.

New technologies of the late nineteenth century had a profound effect on female identity in that, by the close of the century, the discourse about women in the workforce suggested that the woman *is* the machine. As in the case of Bosanquet, women in the workforce often became automatic writers, but many were unable to attain agency as secretaries, telephone operators, or telegraph workers, but instead passively copied down someone else's words or transmitted someone else's voice. The girl in James's 'In the Cage' battles for agency within a workforce that expected and promoted her anonymity and her quiescence. Significantly, the girl is anonymous, suggesting the very identitylessness of her function and the possibility that she is like all of the other nameless hordes of telegraphers, automatically and mechanically transmitting messages 10 hours a day, 6 days a week. Jill Galvan suggests that female mediums were 'exemplary go-betweens because they potentially combined the right kind of presence with the right kind of absence'; to an extent, the girl fits this profile.[66] Yet despite this, she attempts to remedy her non-identity by turning to the heightened powers she believes she possesses. Lisa Gitelman points to the coincidence of writing acting as 'both a psychophenomenon and as typing', believing this 'points doubly to the openness of the word "automatic" during the 1890s'.[67] Automaticity in the nineteenth century thus evoked images of typewriters and telegraphs, spiritualist

---

(1995), 284–307. Savoy is interested in the ghostly, but rather than focusing on uncanny technologies of writing, he suggests that the girl's 'spectral' and 'speculative' knowledge links her to 'the regime of the closet' (p. 286) and thus to the 'text's fundamental queerness' (p. 287).

[66]   Galvan, Jill, *The Sympathetic Medium: Feminine Channeling, the Occult, and Communication Technologies, 1859–1919* (Ithaca: Cornell University Press, 2010), p. 12.

[67]   Gitelman, p. 186.

séances, and the *women* behind all of these operations. Bette London, Alex Owen, and Marlene Tromp have argued that unlike automatic writers like secretaries and telegraphers, spiritualist automatic writers found an empowerment in their status as mediums, giving them sexual and social freedoms that would otherwise have been denied them as women. In passages of automatic writing, mediums express illicit desire, conjuring in their writing a sort of 'double life': 'to be a medium one had to become Other to oneself [...]. The literature of mediumship is thus filled with allusions to the "double life" of its exponents'.[68] The girl in 'In the Cage' is herself a medium, able to receive and transmit telegrams, but also, as she believes, a medium for the thoughts of the socialites she serves.

Although the girl develops a 'double life' (p. 152) – on the one hand bodily action, on the other mental insight – and gains some agency, the empowerment of the spiritualist medium is ultimately denied her. Her telepathy gives her the impression that she can somehow effect change, that she can move upwards in the world and out of the cage by winning Everard's love. This half of her double life, however, is not 'the harmless pleasure of knowing' (p. 198), nor does it liberate her in the way spiritualist mediums had been liberated during séances in which they could take on more sexually confident roles. The girl's telepathic powers finally imprison her in the cage, for these thoughts are a product of the very life she believes they free her from. Because her work is so stultifying, because she is so irrevocably closed off from the classes she admires, and because she is married to Mr. Mudge, whom she decidedly does not love, she must imagine another life for herself. Her mistake lies in her belief that her imaginary life can, in fact, somehow carry her above and through the bars of the social cage. Because she is a telegrapher, she must imagine that she knows 'stories and meaning without end' in order to find at least small satisfaction in an otherwise wholly unsatisfactory existence (p. 155). Her mind automatically creates a diversion for her – her very escape mechanism is a technology of the mind to pacify (and at the same time verify) her automatic existence.

The automatic nature of fin-de-siècle technologies used in the workplace led to a mechanization of human endeavour, which in 'In the Cage' leads to a terrifying crumbling of the self. At the end of the story, the girl realizes at last that she knows and effects nothing in the lives she touches, and that her hopes to become a lady who receives telegrams, instead of being a transmitter of them, are only dreams. To become too involved in the invisible wires connecting minds in 'In the Cage', and to allow oneself to become automated, is to uncannily be dispersed amongst the wires of the cage and the telegraph poles and to have one's identity, finally, mechanized.

The girl's fragmented identity by the end of the story is, again, suggestive of James's own fears about the dangers of the medium: what happens when information is filtered through someone else? Andrew J. Moody discusses nineteenth-century anxieties about the telegraph office, which could dangerously

---

[68] London, p. 129.

expose the secrets of its clients. Moody suggests that 'In the Cage' is ultimately about James's concerns about protecting his private life.[69] But James is also concerned about how writing becomes the site of competing claims: for James, technologies intensify the haunted nature of writing, offering the possibility that outside influence, even of a mechanical nature, can leave a ghostly impression of itself on the final picture.

Postscript

By the end of 1915, Henry James was dying. Even on his deathbed, however, he insisted that Bosanquet take dictation, despite the fact that much of what he said was incoherent. Leon Edel writes that '[i]t is to be noted (Mrs. W.J. reported) that even after [James] lapsed into a coma, his hands continued to move across the bedsheets as if he were writing'.[70] Writing with the ghost of a pen, on the ghost of paper, even at the last James was fascinated by the uncanny possibilities of writing.

---

[69]   Andrew J. Moody, '"The Harmless Pleasure of Knowing": Privacy in the Telegraph Office and Henry James's "In the Cage"', *Henry James Review*, 16 (1995), 53–65. See also Jill Galvan, who discusses anxieties about privacy of information in relation to spiritualist mediums and women working with communication technologies.

[70]   Henry James, *The Complete Notebooks of Henry James*, ed. by Leon Edel and Lyall H. Powers (Oxford: Oxford University Press, 1987), p. 582.

Chapter 2

Sensitive to the Invisible: Photography and the Supernatural in the Holmes Stories, Arthur Conan Doyle's Spiritualism, and Francis Galton's Composite Portraits

Introduction

> [Sherlock Holmes] loved to lie in the very centre of five millions of people, with his filaments stretching out and running through them, responsive to every little rumour or suspicion of unsolved crime.[1]

In this passage, taken from 'The Adventure of the Cardboard Box' (1893), Sherlock Holmes is portrayed as a maze of telegraph wires, his 'filaments' receiving and transmitting messages of criminality across London. He becomes a kind of network for the interception of crimes and criminals, which evokes not only the telegraph wires in James's 'In the Cage' but also the network within the body and, particularly, the brain: filaments are like the electrical currents sending impulses along the nerve centres, and Holmes's body is sparked, or charged, when he connects to the criminals, stimulating him to swift and decisive action. That filaments can be both invisible and material (like the fibres of a spider's web) is also suggestive about the possibility that Holmes is invisibly, yet tangibly connected to the criminal underworld. Significantly, Holmes's filaments are reminiscent of descriptions of his arch-nemesis, Professor Moriarty, who 'sits motionless, like a spider in the center of its web, but that web has a thousand radiations, and he knows well every quiver of each of them'.[2] Like Moriarty, Holmes does not simply transmit impulses along his filaments, but receives them as well, suggesting he is charged with the same energy off which the criminals feed. Furthermore, Holmes revels in the criminal energy, claiming to Watson that '[m]y horror at [Moriarty's] crimes was lost in my admiration at his skill'.[3]

Holmes can telegraphically, telepathically detect the criminals in London, just as he reads Watson's thoughts in Baker Street a few paragraphs later.

---

[1] Arthur Conan Doyle, 'The Adventure of the Cardboard Box', in *The Penguin Complete Sherlock Holmes*, preface by Christopher Morley (Harmondsworth: Penguin, 1981), pp. 888–901 (p. 888).

[2] Doyle, 'The Final Problem', in ibid., pp. 469–80 (p. 471).

[3] Ibid., p. 471.

Although Holmes dismisses his 'small essay in thought reading' as a matter of drawing rational scientific conclusions, he is nevertheless in surprising proximity to the unempirical and unsavoury: Holmes's filaments connect him not only to criminality, but to the uncanny power to read minds.[4] Watson is anxious about this possibility, however, arguing that although Holmes's deductive abilities would appear to the 'uninitiated' as those of a 'necromancer', he is actually 'the most perfect reasoning and observing machine that the world has ever seen'.[5]

The notion of Holmes as an 'observing machine' also links him to a contemporary 'observing machine', the camera.[6] Indeed, Doyle's 'The Adventure of the Cardboard Box' evokes the notion of the camera: from 1842 the term 'box camera' was used to describe a hand-held camera shaped like a box, and in the 1880s, a type of box camera known as a 'detective camera' was 'adapted for taking instantaneous photographs' (OED). Ronald R. Thomas has already made connections between Holmes and the contemporary discourse on photography, suggesting, as I do here, that we can think of both the camera and Holmes as 'the literary embodiment of the elaborate network of visual technologies that revolutionized the art of seeing in the nineteenth century'.[7] For Thomas, visual technologies like the camera changed the way Victorians thought about seeing because they could capture what the eye could not – they could make the invisible visible. As Kate Flint suggests, developments in scientific fields like physiognomy and the use of technologies like the microscope meant that Victorian ways of seeing were heavily influenced by ideas about what was undetectable to the naked eye.[8] Furthermore, Thomas suggests that both the

---

[4]   Doyle, 'Cardboard Box', p. 890.

[5]   Doyle, 'A Scandal in Bohemia', in *Complete Sherlock Holmes*, pp. 161–75 (p. 161); *A Study in Scarlet*, in ibid., pp. 15–86 (p. 23).

[6]   Significantly, in *Telegraphic Realism: Victorian Fiction and Other Information Systems* (Stanford: Stanford University Press, 2008), Richard Menke argues that 'photography and telegraphy were also linked by both contemporary discourse and their intertwined histories' (p. 137), suggesting a further connection between Holmes and photography.

[7]   Ronald R. Thomas, 'Making Darkness Visible: Capturing the Criminal and Observing the Law in Victorian Photography and Detective Fiction', in *Victorian Literature and the Victorian Visual Imagination*, ed. by Carol T. Christ and John O. Jordan (Berkeley: University of California Press, 1995), pp. 134–68 (p. 135).

[8]   According to Flint, looking at what was invisible was crucial to understanding the ways in which Victorians saw, particularly with developments in scientific fields like physiognomy and the use of technologies like the microscope. See Kate Flint, *The Victorians and the Visual Imagination* (Cambridge: Cambridge University Press, 2000), p. 8. While these innovations in science seemed to suggest that vision could be mapped and quantified, they also highlighted '[t]he slipperiness of the borderline between the visible and the invisible, and the question which it throws up about subjectivity, perception and point of view' (Flint, p. 2). Indeed, as Jonathan Crary argues, vision in the nineteenth century was marked by disruption, discontinuity, and disorder – vision had gained a 'new autonomy and abstraction'. See Jonathan Crary, *Techniques of the Observer: On Vision and Modernity in the Nineteenth Century* (Cambridge, MA: MIT Press, 1998), p. 14.

camera and the literary detective 'developed a practical procedure to accomplish what the new discipline of criminal anthropology attempted more theoretically: to make darkness visible – giving us a means to recognize the criminal in our midst by changing the way we see and redefining what is important for us to notice'.[9] But while Thomas is interested in the ways in which the camera and literary detectives like Holmes observe and capture the criminal, I suggest that Holmes's 'virtually photographic powers' mean he can be linked to the invisible in a different way:[10] despite assumptions that Holmes and photography were the epitome of ruthless, rigorous figuring, they were nevertheless sensitive to invisible elements normally undetectable to rational perception, such as the supernatural.[11] Furthermore, a close reading of Doyle's Holmes stories shows that Holmes's sensitivity to spiritualist insight and connections to hypnotic trance states actually inspire and perfect his deductive abilities. Holmes is, as the quote beginning this chapter suggests, 'responsive to every little rumour or suspicion of unsolved crime': he is receptive to suspicions of crime just as a photographic plate is sensitive to light.[12]

---

[9] Thomas, 'Making Darkness', p. 135.

[10] Ibid., p. 134.

[11] Critics have argued that Sherlock Holmes's rational, scientific identity ultimately prohibits him from association with either supernatural or criminal elements. See, for example, James Kissane and John M. Kissane, 'Sherlock Holmes and the Ritual of Reason', *Nineteenth-Century Fiction*, 17 (1963), 353–62 (p. 361). The Kissanes argue that in the Holmes stories, reason is a kind of ritual which is crucial to the triumph of the detective over the criminal. In Jasmine Yong Hall's 'Ordering the Sensational: Sherlock Holmes and the Female Gothic', *Studies in Short Fiction*, 28 (1991), 295–303, Hall argues that 'both the Gothic elements and the female clients in these stories play an important role in establishing the rational detective as a powerful, patriarchal hero' (p. 295). Although Jesse Oak Taylor-Ide's 'Ritual and the Liminality of Sherlock Holmes in *The Sign of Four* and *The Hound of the Baskervilles*', *ELT*, 48 (2005), 55–70, suggests that Holmes is associated with the liminal, she ultimately sees him as the embodiment of the power of reason.

[12] See also Srdjan Smajic, *Ghost-Seers, Detectives, and Spiritualists: Theories of Vision in Victorian Literature and Science* (Cambridge: Cambridge University Press, 2010) for an examination of Holmes's links to spiritualism and ghosts. Because Smajic's work was published just as mine was going to press, it has not been possible to fully engage with it. While both our works explore connections between the Sherlock Holmes stories and the supernatural, and the slippages between genres and Victorian science, our emphases are different. Smajic articulates a 'persistent metatextual concern in detective fiction; the anxiety that generic purity is unattainable; that the supposedly rational genre in which the supposedly rational Holmes feels at home is everywhere contaminated by the supernatural, occult or irrational' (p. 3). He investigates the 'hybridization' between ghost and detective fiction, suggesting that this is a result of new sciences on seeing in the Victorian period, particularly in terms of contemporary interest in the invisible (p. 7). My focus is on how the Holmes stories are connected with late nineteenth-century concerns about writing, new technology, and the borderline between science and the supernatural. Furthermore, I focus on the discourse surrounding Victorian photography, particularly spirit photography, and Doyle's interest in spiritualism as a means of discussing Holmes's links with the supernatural.

Although the camera was, as Jonathan Crary suggests, 'a metaphor for the most rational possibilities of a perceiver within the increasingly dynamic disorder of the world',[13] invisible elements were still materializing in its frames – the rise of spiritualism and spirit photography meant that Victorians were using the camera to capture ghosts.[14] Arthur Conan Doyle had been interested in spiritualism since 1886, and the passage beginning this chapter reflects both the language and imagery of late Victorian spiritualist practice.[15] The filaments Holmes extends are suggestive of the spiritualist séances in which mediums secreted fluid matter, often described as filament-like and which was identified by Charles Richet in 1923 as ectoplasm. In volume 2 of *The History of Spiritualism* (1926), Doyle records an account of this ectoplasmic formation: '[w]e find, too, in *The Spiritualist* that while the materialized spirit Katie King was manifesting herself through Florence Cook, '[s]he was connected with the medium by cloudy, faintly luminous threads'.[16] Richet describes ectoplasm in the following terms: '[t]his ectoplasm makes *personal* movements. It creeps, rises from the ground, and puts forth tentacles like an amoeba. It is not always connected with the body of the medium but usually emanates from her, and is connected with her'.[17]

Significantly, the notion of material flowing from the body and language like 'filament' is also suggestive of the discourse surrounding spirit photography, which became popular in the 1860s when William Mumler of Boston produced photographs in which ghostly figures and balls of light appeared. Dozens of theories were circulated about how the camera could capture invisible beings, but many spiritualists believed that the end result had something to do with a fluid or psychic force captured by the camera's sensitive plate. For example, Doyle argued

---

[13] Crary, *Techniques*, p. 53.

[14] X-rays (invented in 1895) were other invisible elements which brought up anxieties about the dangers of invisible forces: x-rays could magically capture the interior of the body, portraying ghostly images of the skeleton, but they also seemed to have the potential to invade the body, sparking concern about the effects of such an intrusion. See Allen W. Grove, 'Röntgen's Ghosts: Photography, X-Rays, and the Victorian Imagination', *Literature and Medicine*, 16 (1997), 141–73.

[15] Doyle insists in *The History of Spiritualism*, 2 vols (London: Cassell, 1926), II, that '[i]t has been said ... that the author's advocacy of the subject [spiritualism], as well as that of his distinguished friend Sir Oliver Lodge, was due to the fact that each of them had a son killed in the war, the inference being that grief had lessened their critical faculties and made them believe what in normal times they would not have believed. The author has many times refuted this clumsy lie, and pointed out the fact that his investigation dates back as far as 1886' (p. 224).

[16] Ibid., p. 90.

[17] Charles Richet, *Thirty Years of Psychical Research: Being a Treatise on Metaphysics*, trans. by Stanley de Brath (London: Collins, 1923), p. 523. Although the term 'ectoplasm' was not adopted until the early twentieth century, materializations during séances had been documented as a regular occurrence since the 1870s, and both Doyle and Richet suggest that these were examples of ectoplasm.

that 'the effect is produced by a sort of ray carrying a picture upon it which can penetrate solids, such as the wall of a dark slide, and imprint its effect upon the plate'.[18] He believed that ectoplasm was another effluvium which could penetrate the camera's plate: 'ectoplasm once formed can be moulded by the mind' and can thus be imprinted onto the photographic plate.[19] The spiritualist magazine *Light* suggested that spirit photographs were created with an invisible agency:

> [n]ow, suppose some entity invisible to us, because of our limited vision is able to start vibrations in some way beyond the violet end of the spectrum; these vibrations striking the sensitised plate, might become less rapid or 'be degraded' and 'fluoresce', and so come into the arctic range and be photographed.[20]

Theories of imponderable fluids, Janet Oppenheim argues,

> were useful to the spiritualist, as to the psychical researcher, seeking to explain thought transference, for as it flowed from one person to another through the nervous system, the liquid might convey the thoughts of the first to the second. It might even, through obscure interactions between animate and inanimate, provoke rappings and movements of furniture.[21]

In particular she credits Karl von Reichenbach's theory of the odic force, another imponderable fluid which was believed to effect psychic phenomena, as highly influential on the researches of the SPR.[22]

While Oppenheim primarily links ideas about psychic fluids to mesmerism (a subject which I will return to later in this chapter and in Chapter 3), the notion of a powerful effluvium in connection with spirit photography, something invisible which could somehow materialize visibly on the sensitive plate of the camera, is telling about the paradoxes within conceptions of photography in the Victorian period. How could photography, commonly associated with realism, and in particular with the rational and evidential nature of police work, also be used to prove the existence of ghosts?[23] James R. Ryan has suggested that 'despite the common assumption that it was a truthful means of representing the world, photography was also a social practice whose meanings were structured

---

[18] Doyle, *History*, II, p. 145.

[19] Ibid., p. 117.

[20] 'Spirit Photography', *Light*, 25 November 1893, p. 562.

[21] Janet Oppenheim, *The Other World: Spiritualism and Psychical Research in England, 1850–1914* (Cambridge: Cambridge University Press, 1985), pp. 218–19.

[22] Ibid., p. 219.

[23] See, for example, Lorraine Daston and Peter Galison, 'The Image of Objectivity', *Representations*, 40 (1992), 81–128, and, John Tagg, *The Burden of Representation: Essays on Photographies and Histories* (London: Macmillan, 1988), who have argued that Victorians associated photography with truth and infallible empirical evidence.

through cultural codes and conventions'.[24] In this chapter, I would like to examine photography through the 'cultural codes and conventions' of Victorian spiritualism and science to suggest that perceptions about the flawless visual powers of the camera actually reflect wider anxieties about policing the boundaries between science and supernaturalism. While Jennifer Tucker has discussed the notion that 'Victorians did not, in fact, accept photographic evidence as unconditionally true', I suggest that they deeply desired this to be the case.[25] Daniel Novak's theory that Victorian 'photographic realism produces and records that which was never there in the first place – "ideal pictures" of the imagined, the impossible, and the fictional' is particularly useful here.[26] Scientists and spiritualists alike hoped that photographic proof would make their researches unassailable; in fact, the photographic evidence they provided often undermined their work. This chapter begins with case studies of Arthur Conan Doyle and Francis Galton to demonstrate how the discourse on photography revelled in the shared energies of science and the supernatural at the end of the century. Galton, cousin to Charles Darwin, was the founder of eugenics, invented the fingerprinting system in 1888, and made advances in various fields such as meteorology, psychology, and anthropology. While he attempts to use photography to make empirical claims about criminality, he is ultimately unable to police ghostly elements from his studies. In Doyle's case, he is haunted by his impossible desire to capture spirits within the sensitive plate of the camera, attempting to rigorously control and constrain the inchoate elements of the spirit world. Indeed, Doyle's works are themselves marked by a similar confusion between the rationality and order of the Sherlock Holmes adventures and the provisionality of the spiritualist writings. This is an opposition which proves to be superficial, however: not only does Holmes himself have supernatural qualities, but Doyle's spiritualist writings demonstrate a desire to incorporate both evidential proof and scientific tests sympathetic to spiritualist belief.

---

[24] James R. Ryan, *Picturing Empire: Photography and the Visualization of the British Empire* (Chicago: University of Chicago Press, 1997), p. 17. See also Allan Sekula, 'On the Invention of Photographic Meaning', in *Photography in Print: Writings from 1816 to the Present*, ed. by Vicki Goldberg (New York: Simon and Schuster, 1981), pp. 452–73, who argues that 'if we accept the fundamental premise that information is the outcome of a culturally determined relationship, then we can no longer ascribe an intrinsic or universal meaning to the photographic image' (p. 454).

[25] Jennifer Tucker, 'Photography as Witness, Detective and Impostor: Visual Representations in Victorian Science', in *Victorian Science in Context*, ed. by Bernard Lightman (Chicago: University of Chicago Press, 1997), pp. 378–408 (p. 380).

[26] Daniel Novak, *Realism, Photography, and Nineteenth-Century Fiction* (Cambridge: Cambridge University Press, 2008), pp. 38–9.

Doyle's Spirit Photography

In the 1890s, Doyle became interested in optics and travelled to Vienna in order to study in that field. He opened an eye practice in London, but with the success of the Sherlock Holmes stories he gave up the business, sold it, and used the proceeds to buy a camera. His interest in photography seems to date from the 1880s, feeding into the period in which he developed an interest in spiritualism. From 1881 to 1885, Doyle submitted a number of essays on photography to the *British Journal of Photography*. In the 1890s, he compiled albums of photographs, although unfortunately, none of his own photographs remain today.[27] He was fascinated primarily by the mechanism of the camera itself, and his writings on the subject are preoccupied with the chemical changes that work to create a photograph. His essays are, for the most part, technical discussions of the photographic process; they describe the type of camera he used, the length of exposure, and the best way to create a makeshift darkroom.[28]

In his essay 'After Cormorants with a Camera' (1881), his first photographic essay to be published in the *British Journal of Photography*, Doyle even uses his camera to solve a mystery. While hunting on the Isle of May in the Firth of Forth, Doyle and his friends hire a guide, an elderly man they facetiously name Sinbad. When he fails to bring their lunch at the appointed time, Doyle and his friends seek him out and watch him unnoticed, as he 'had just emptied a third of our whisky bottle down his throat, and was bending down at the little stream while he filled it up with water'.[29] Doyle captures him on camera, and the unlucky guide is humiliated when he is presented with the photograph proving his guilt – the ruse is up, Doyle has caught his criminal with the ultimate detective, the camera. Significantly, in 'The Adventure of the Veiled Lodger' (1927), Watson mentions one of Holmes's adventures which has not yet been shared with the public, a case 'concerning the politician, the lighthouse, and the trained cormorant'.[30] This case might refer to Doyle's essay (which recounts how Doyle and his friends are hunting cormorants on an island with a lighthouse), suggesting that Doyle is playing with the notion that Holmes is a kind of camera who seeks out and captures the criminal.

By the 1920s, however, Doyle was less concerned with the camera's technical workings or its role in solving crime, and more interested in how the camera could solve a different kind of mystery – that of proving the existence

---

[27] See John Michael Gibson and Richard Lancelyn Green, eds, *The Unknown Conan Doyle: Essays on Photography* (London: Secker & Warburg, 1982), pp. x–xix.

[28] Two of the photographic essays, 'A Few Technical Hints', in ibid., pp. 38–9, and 'Trial of Burton's Emulsion Process', in ibid., pp. 40–42, focus solely on the chemical processes involved in making a photograph. They are complicated and technical, demonstrating that Doyle had an excellent knowledge of amateur photography.

[29] Doyle, 'After Cormorants with a Camera', in ibid., pp. 1–12 (p. 9).

[30] Doyle, 'Adventure of The Veiled Lodger', in *Complete Sherlock Holmes*, pp. 1095–1102 (p. 1095).

of spirits. His writings were marked by a seemingly similar contradiction between the logic and reason of the Holmes stories and the supernaturalism of his spiritualism texts. Surprisingly, the Holmes stories contain scenes which would not be out of place in spiritualism tracts and are particularly reminiscent of Doyle's descriptions of spirit photography. In 'The Man with the Twisted Lip' (1891), for example, the opium den might equally describe a séance room or a spirit photograph:

> Through the gloom one could dimly catch a glimpse of bodies lying in strange fantastic poses, bowed shoulders, bent knees, heads thrown back, and chins pointing upward, with here and there a dark, lack-lustre eye turned upon the newcomer. Out of the black shadows there glimmered little red circles of light, now bright, now faint, as the burning poison waxed or waned in the bowls of the metal pipes.[31]

The strange 'poses' of the opium-eaters (including Holmes, who is lurking in the frame) are like the poses of the 'fantastic and grotesque' spirits in a spirit photograph Doyle describes in *The Vital Message* (1919) – here, luminous arms and heads materialize on the photographic plate.[32] Although 'little red circles of light, now bright, now faint' refers to the opium pipes, it also evokes spirit lights and the significance of red lighting to psychic investigation. According to John Harvey, '[r]ed is a spectral wavelength (in more senses than one), which, during the nineteenth and early twentieth centuries, was believed to be conducive to concentrating the apparitional image both inside the shadowy chamber of the medium's cabinet and onto the camera's dark slide'.[33]

In *The Sign of Four* (1890) Watson also describes what could be a darkened séance room, where spirits are flitting about the room:[34]

> There was, to my mind, something eerie and ghost-like in the endless procession of faces which flitted across these narrow bars of light, – sad faces and glad, haggard and merry. Like all human kind, they flitted from the gloom into the light, and so back into the gloom once more. I am not subject to impressions, but the dull, heavy evening, with the strange business upon which we were engaged, combined to make me nervous and depressed. I could see from Miss Morstan's manner that she was suffering from the same feeling. Holmes alone could rise superior to petty influences.[35]

---

[31]   Doyle, 'The Man with the Twisted Lip', in ibid., pp. 229–44 (p. 231).

[32]   Doyle, *The Vital Message* (New York: George H. Doran, 1919), p. 141.

[33]   John Harvey, *Photography and Spirit* (London: Reaktion Books, 2007), p. 26.

[34]   See, for example, Doyle's *Vital Message*, in which he gives the psychic investigator William Crookes's description of a séance: 'I have seen luminous points of light darting about, sitting on the heads of different persons' (p. 115).

[35]   Doyle, *The Sign of Four*, in *Complete Sherlock Holmes*, pp. 89–158 (p. 98).

The faces Watson describes are suggestive of the spirit faces Doyle later sees in the photographs taken by the psychic investigator Dr. Gustave Geley: 'the faces are, on the whole, pretty and piquant, though of a rather worldly and unrefined type'.[36] Furthermore, when Watson insists he is not 'subject to impressions', he might be referring to a photographic impression, suggesting that while he is not often exposed to photographs, Holmes is. In this reading, Holmes might be another one of the ghostly faces that materializes in the gloom, just as the spirits do in a spirit photograph. Indeed, Holmes is also 'subject to impressions'; he is exposed to ideas and feelings, but these are not the 'petty influences' of nervousness and depression. Instead, he seems to be subject to higher influences (he can 'rise superior'), suggesting not only his elevated reasoning abilities, but also his link to 'higher communications from the other side', articulated in spiritualism.[37]

Just as Doyle's detective and spiritualism writings seemed to share a script, his early and later photographic interests were never as separate as they seem and, despite himself, distinctions between science and the supernatural in Doyle's photography and spiritualism writings are difficult to ascertain.[38] Even his technical essays, published in the early 1880s, demonstrate an interest in the connections between photography and death, perhaps foreshadowing his later interests in photography's link to the spirit world. In 'After Cormorants with a Camera', the same essay in which he exhorts the evidential nature of the camera, Doyle expresses his desire to photograph death, not merely in its finality, but as it is happening; the process of death, the moment between life and death. He describes it as a 'photographic novelty', and asks 'why not take a cormorant at the moment of its being shot?'[39] He comments on the plate, writing that it came out 'sharp as a die', suggesting that he is punning on death itself, which is here clarifying and illuminating.[40] The photograph (which we will never see) lends itself to a complicated interpretation, for while we can imagine a photograph of a cormorant as it is shot mid-flight, it is difficult to say at what moment we have captured the cormorant. Has the bullet hit it yet? Is it dead? Or is it only nearly dead? When is the moment of the last heartbeat? Significantly, the term 'shooting' in Doyle's essay refers both to hunting and to taking a picture, which is suggestive about the connections between photography and death.[41] Likewise, in 'On the Slave Coast with a Camera' (1882), the word 'take' means dually to capture and kill men in

---

[36] Doyle, *Vital Message*, p. 144.

[37] Ibid., p. 117.

[38] See also Srdjan Smajic, who is interested in the ways in which detective fiction and ghost story genres are blurred.

[39] Doyle, 'After Cormorants', p. 10.

[40] Ibid., p. 11.

[41] While the OED gives 1890 as the date in which the photographic definition of the word was in circulation, Doyle's language suggests he was referring to both hunting and taking a picture.

combat, and also to 'take' a photograph.[42] In this essay, an African chief boasts to Doyle that in 'the last campaign [he] had taken five hundred men', to which Doyle responds that 'he could take as many as that in a single moment'.[43] In his anecdote, Doyle recounts how the chief was frightened of him afterwards, and how many of the natives would jump from their boats when they saw his camera pointed at them, believing it was a gun.[44]

Just as photographic words like 'take' and 'shoot' elude a single definition, distinctions between photography's technical workings and its supernatural power are also ambiguous in the nineteenth-century conception of photography. As Lindsay Smith suggests, 'from its inception, photography held a connection with magic'.[45] Furthermore, Geoffrey Batchen points out that the literature on photography treated the process as a kind of necromancy, fuelling the myth that the camera had the ability to strip away layers of the soul with each successive photograph.[46] Doyle's writings about photography give evidence of his anxiety about the ambiguity between the technological advances of science and the inexplicable and unsystematic nature of the supernatural: he knows the technical terms, the correct chemicals to use, and the reactions that must occur for a photograph to be developed, and yet he still refers to photography as the 'black art', and in his spiritualist writings discusses how the camera is an instrument capable of capturing images of the dead.[47]

For Doyle, the notion that photography escapes systematization proved particularly problematic, especially since he wanted to use the camera to somehow regulate the spirit world. By visibly demonstrating their existence in the photographs, Doyle could safeguard the spirits from the scepticism of the scientific community. However, he was also attempting to capture invisible beings through a medium which was designed to capture the visible. Doyle's photographs are haunted, then, not only by the spirits who materialize on the plate but also by the notion that these images elude capture, and that his efforts to contain the supernatural are futile. Nevertheless, Doyle insisted that photographs of spirits offered insurmountable evidence of the existence of a spirit world. His interest in spirit photography was so strong that when the Society for the Study

---

[42]   Doyle, 'On the Slave Coast with a Camera', in *The Unknown Conan Doyle*, pp. 13–22. According to the OED, 'take' has been used in photographic language since 1859.

[43]   Ibid., p. 19.

[44]   For a discussion of Doyle and imperialism, see, for example, Catherine Wynne, *The Colonial Conan Doyle: British Imperialism, Irish Nationalism, and the Gothic* (London: Greenwood, 2002).

[45]   Lindsay Smith, *The Politics of Focus: Women, Children and Nineteenth Century Photography* (Manchester: Manchester University Press, 1998), p. 2.

[46]   Geoffrey Batchen, *Burning with Desire: The Conception of Photography* (Cambridge, MA: MIT Press, 1997), pp. 208–11.

[47]   Doyle, 'To the Waterford Coast and Along It', in *The Unknown Conan Doyle*, pp. 51–9 (p. 53).

of Supernormal Pictures (SSSP) formed in 1918 he became its vice-president. He also wrote on the subject in *The Case for Spirit Photography* and *The Coming of the Fairies*, both published in 1922, and in the second volume of *The History of Spiritualism*.

In *The Case for Spirit Photography* Doyle insists that he himself has been successful in developing plates which depict supernatural phenomena, a fact which he suggests makes their authenticity even more perfect.[48] The pamphlet is filled with photographs in which indistinct blurs of ectoplasm hover near the sitters for the photographs, and bodiless heads loom. Doyle's second volume of *The History of Spiritualism* marks spirit photography as an important, technologically advanced method for supplying evidence that the world of spirits is a real one. *The Coming of the Fairies* is an account of the Cottingley fairy incident, in which two young girls, Elsie and Frances Wright, took photographs of themselves posed with fairies.[49] The photographs were the subject of great controversy, and Doyle was heavily criticized for his willingness to believe the evidence of a few photographs taken by children. Doyle first wrote about the incident in *The Strand* in the December issue of 1920, using the magazine he usually reserved to uphold 'the science of deduction' in the figure of Holmes to declare his belief that the photographs were authentic, and that fairies were real.[50]

Doyle believed that the camera was the best means of giving evidence, not only because it visibly offered up the fairies, but also because the technology of the camera potentially still held unknown mysteries which could be sensitive to hitherto unseen forces. Doyle suggests that the fairies are 'separated from ourselves by some difference of vibrations. We see objects within the limits which make up our colour spectrum, with infinite vibrations, unused by us, on either side of them'.[51] For Doyle, the 'sensitive plate' of the camera was a means of recording beings like spirits and fairies because the potential of the development process was not fully understood.[52]

Doyle's uncertainty about the limits of photographic technology suggests his ambiguity about the union of science and the supernatural. While he believes that scientific interference in spiritual phenomena is invasive and futile, he is also deeply concerned with proving the realities of spiritualism. He dismisses the methods of the SPR, but argues that a different scientific methodology, more sympathetic to spiritualism, needs to be implemented in order to come to a greater understanding of supernatural phenomena. For example, Doyle writes, '[s]peaking generally, it may be said that the attitude of organized science during these thirty years was as

---

[48] Doyle, *The Case for Spirit Photography* (London: Hutchinson, 1923), pp. 18–19.

[49] Doyle, *The Coming of the Fairies* (New York: Weiser, 1979).

[50] Chapters bearing the title 'The Science of Deduction' appear in *A Study in Scarlet* and in *The Sign of Four*. 'The Science of Deduction' is also mentioned in an article written by Holmes in *A Study in Scarlet*.

[51] Doyle, *Fairies*, pp. 13–14.

[52] Ibid., p. 56.

unreasonable and unscientific as that of Galileo's cardinals, and that if there had been a Scientific Inquisition, it would have brought its terrors to bear upon the new knowledge'.[53] Doyle argues that testing should be modified to suit the conditions of the séance, just as any scientific test should be designed effectively for the conditions of the experiment:

> [i]f a small piece of metal may upset a whole magnetic installation, so a strong adverse psychic current may ruin a psychic circle. It is for this reason, and not on account of any superior credulity, that practising Spiritualists continually get such results as are never attained by mere researchers.[54]

Doyle is deeply anxious about the infiltration of scientific investigators into spiritualism, not only because he fears their methods interrupt and hinder spiritual phenomena, but more important, because they might somehow strip away the fascinating possibilities of the unknown:

> Victorian science would have left the world hard and clean and bare, like a landscape in the moon; but this science is in truth but a little light in the darkness, and outside that limited circle of definite knowledge we see the loom and shadow of gigantic and fantastic possibilities around us, throwing themselves continually across our consciousness in such ways that it is difficult to ignore them.[55]

His description of the world of science surrounded by looming unknown possibilities is reminiscent of William James's overgrown terrace of the mind, in which 'uncouth forms lurk in the shadows'.[56] While I argue in Chapter 1 that the Gothic nature of mental science was both inspiring and threatening for writers like Henry James, the threat for Doyle is that the world of the unknown may be unable to materialize because of the scepticism and methods of the scientific community. Furthermore, although science itself threatens to strip and sterilize the inspirational possibilities of the supernatural, Doyle needs the support of scientific methods to demonstrate that the spiritual phenomena are authentic.

Doyle is searching for a moment in which science and the supernatural might share a language, but he is also anxious about the repercussions of such a collision. Possibly, Doyle finds this moment in the figure of Holmes, the fictional character that, significantly, he was most anxious about (he disliked writing about him and unsuccessfully tried to kill him off in 'The Final Problem' [1893]). After all, when Doyle writes that science is but a 'little light in the darkness', it seems that he is also referring to Holmes. The theme of the 'Light in the Darkness' that Holmes can shed is constant throughout the short stories and the long: Watson continually asks

---

[53] Doyle, *History*, I, p. 185.

[54] Ibid., II, p. 318.

[55] Doyle, *Fairies*, p. 125.

[56] William James, 'Frederic Myers's Service to Psychology', in *A William James Reader*, ed. by Gay Wilson Allen (Boston: Houghton Mifflin, 1971), pp. 155–64, p. 56.

if Holmes can throw no light onto the darkness of the case, as do many of Holmes's clients.[57] But if Holmes does shed light on the darkness, will his logical reasoning banish the 'gigantic and fantastic possibilities all around us'? On the contrary, '[Holmes] refused to associate himself with any investigation which did not tend towards the unusual, and even the fantastic'.[58] Furthermore, as I will show in the final section of this chapter, Holmes's powers of deduction are actually enhanced by insight gained through trance states and his connections to the spirit world.

Francis Galton's Composite Portraits: Criminals and Ghosts

Photography was integral to Victorian detective work, which relied heavily on the technology as a means of identifying and categorizing criminals. As Jennifer Green-Lewis points out, '[t]he mug shot was conceived almost as soon as knowledge of daguerreotypy spread', suggesting that compiling and memorizing criminal faces was the necessary key to successful criminal capture.[59] By the 1880s the Bertillon system, which involved identification of a person by body measurements and photographs, had been put into practice. Holmes himself was familiar with it, expressing 'enthusiastic admiration' of its inventor, Alphonse Bertillon, in 'The Adventure of the Naval Treaty' (1893).[60] Although the Bertillon system was eventually replaced by fingerprinting, Francis Galton (who invented the fingerprinting system) was photographed as part of Bertillon's collections of mug shots in 1893.[61]

From 1877–1885, Galton was fascinated by the ways in which photography, in particular composite portraiture, or composite photography – a new technology he

---

[57]   For example, Chapter 7 in *A Study in Scarlet* is titled 'The Light in the Darkness' (p. 46). The notion of shedding or throwing light on a case is also mentioned in *The Sign of Four*, *The Hound of the Baskervilles*, 'The Boscombe Valley Mystery', 'The Adventure of the Speckled Band', 'The Adventure of the Stockbroker's Clerk', and 'The Adventure of the Naval Treaty'.

[58]   Doyle, 'The Speckled Band', in *Complete Sherlock Holmes*, pp. 257–73 (p. 257).

[59]   Jennifer Green-Lewis, *Framing the Victorians: Photography and the Culture of Realism* (Ithaca: Cornell University Press, 1996), p. 200. The daguerreotype was presented to the public in 1839; the 1840s saw the rise of the mug shot.

[60]   Doyle, 'The Naval Treaty', in *Complete Sherlock Holmes*, pp. 447–69 (p. 460). In *The Hound of the Baskervilles* (*Complete Sherlock Holmes*, pp. 667–766), Holmes is offended when Dr. Mortimer lists him as 'the second highest expert in Europe' after Bertillon (p. 672).

[61]   Galton is usually discussed in connection with criminality and the Sherlock Holmes stories in terms of his fingerprinting system. See, for example, Ronald R. Thomas, 'The Fingerprint of the Foreigner: Colonizing the Criminal Body in 1890s Detective Fiction and Criminal Anthropology', *ELH*, 61 (1994), 655–83, and Gita Panjabe Trelease, 'Time's Hand: Fingerprints, Empire, and Victorian Narratives of Crime', in *Victorian Crime, Madness and Sensation*, ed. by Andrew Maunder and Grace Moore (Aldershot: Ashgate, 2004), pp. 195–206. The links between Galton, criminality, Doyle, and photography, however, are crucial to revealing Holmes's photographic powers, which link him both with the notion of the all-seeing detective and, surprisingly, with criminal and supernatural associations.

had created – could help to regulate and systematize identity, especially criminal identity. Composite photographs were made by superimposing photographs of various faces in order to reveal a single, blended face.[62] According to Galton, 'the process of composite portraiture is one of pictorial statistics',[63] the purpose of which was to mathematically create a *type* of person. Galton was convinced by the scientific necessity and importance of defining types within the human species, and believed that composite photography was the most accurate method of obtaining characteristic humans. He maintained that composites would offer limitless possibilities for the scientific community, and for society at large. They could be used, for example, to 'give typical pictures of different races of men', or to give the most accurate likeness of a single individual, since instead of depicting that person in a single attitude, they would represent a melding of all of that person's attitudes: smiles and frowns would be combined to create a kind of ideal photograph.[64] For the purposes of heredity, the composite photographs could help animal breeders determine what the offspring of any animals would look like, just as husbands and wives could combine their features in a single photograph in order to ascertain how their children would look.

Galton was particularly interested in the ways in which his composites could detect and control the potential for crime, arguing that the composites he made of criminals would represent 'not the criminal, but the man who is liable to fall into crime'.[65] He created 'the criminal' after consulting with Sir Edmund Du Cane, the chairman of the Prison Commission. Du Cane had a large collection of photographs of criminals, which Galton borrowed in order to sort them into 'portraits of criminals convicted of murder, manslaughter, or crimes accompanied by violence', and then to make from them a composite.[66] Du Cane applauded Galton's composites, believing they were a way to find and stop criminals before they turned to crime: '[i]n considering how best to deal with and repress crime, it occurred to me that we ought to try and track it out to its source, and see if we cannot check it there instead of waiting till it has developed and then striking at it'.[67] Significantly, Du Cane's use of the word 'develop' in connection with crime is suggestive not only of the potential threat of crime unfolding, but also of developing a photograph.[68] The camera acts here as a means both of capturing a man before he develops into a fully fledged criminal and, paradoxically, of symbolically creating (or developing) criminal behaviour.

---

[62]   Galton, 'Composite Portraits', *Journal of the Anthropological Institute*, 8 (1878), pp. 132–44 (p. 132).

[63]   Galton, *Generic Images* (London: William Clowes and Son, 1879), p. 5.

[64]   Galton, 'Composite', p. 140.

[65]   Ibid., p. 135.

[66]   Galton, *Generic*, p. 5.

[67]   Galton, 'Composite', p. 142.

[68]   The term 'develop' was first used in connection with photography in 1845 (OED).

In the faces of Galton's composites of criminals, we are meant to see the identifying features of the type of person liable to crime, but find instead puzzlingly bland faces, which could belong to anyone. In light of Du Cane's sinister comments, in which pre-crime is already a crime, the fact that the faces are so nondescript implies that everyone (or no one) is the criminal type. Galton himself states that the composites of the criminals

> represent no man in particular but portray an imaginary figure possessing the average features of any group of men. These ideal faces have a surprising air of reality. Nobody who glanced at one of them for the first time would doubt its being the likeness of a living person, yet, as I have said, it is no such thing; it is the portrait of a type, and not of an individual.[69]

Yet while these composites are meant to depict the criminal with 'mechanical precision', the language Galton uses to describe the criminal is imprecise, emotional, and even literary.[70] The composite of the criminal is an 'ideal', an 'imaginary' image, and far from being the record of an exacting new science designed to track and apprehend the criminal, composite photographs actually invent new and perfect criminals who cannot be captured by the camera because they do not even really exist. Although the composites are designed to define a type of identity, instead they seem to suggest that selfhood is malleable and able to be blended with other selves. The end results of Galton's composites destabilize identity more fully than they encapsulate it: the criminal is the photograph of every man, or any man, and blends of men and women result in 'the production of a face, neither male nor female'.[71] The criminal type is submerged in the unremarkableness of the face, while other composites are asexual; in their very attempt to demonstrate that which can be quantified, they are unquantifiable.

In his examination of the composites, Daniel Novak discusses how Joseph Jacobs, a Jewish social scientist, folklorist, and literary critic, thought that the composites portraying Jewish identity (which he had asked Galton to produce) were 'more ghostly than a ghost, more spiritual than a spirit'.[72] For Jacobs, Novak argues, '[t]he Jewish composites represent at once biological fact and biblical specter, an inherited racial body and a mystical, ghostly inheritance'.[73] In *The Hound of the Baskervilles* (1901), Holmes also connects physically inherited traits to the numinous when he realizes the portrait of Hugo might just as easily be that of Stapleton: 'it is an interesting instance of a throw-back, which appears to be both physical and spiritual. A study of family portraits is enough to convert a man to the doctrine of reincarnation'.[74] Holmes sees Stapleton's face in Hugo's features,

---

[69] Galton, 'Composite', p. 133.

[70] Ibid., pp. 132, 134.

[71] Galton, *Generic*, p. 3.

[72] Novak, p. 102.

[73] Ibid., p. 103.

[74] Doyle, *Hound*, p. 750.

suggesting that concealed beneath the layers of paint are the different members of the Baskerville family, waiting to 'spr[i]ng out from the canvas'.[75] The 'spiritual' emergence of extra faces in the portrait is suggestive not only of spirit photography, but also of Galton's 'ghostly accessories', the blurred and multiplied features that sometimes resulted from superimposing pictures to create the composites.[76] In one composite, Galton combines the photographs of two brothers and one sister in order to obtain the most notable features of that family. He refers to the slightly out-of-focus effect in this particular composite as 'ghosts': '[g]hosts of portions of male and female attire, due to the peculiarities of the separate portraits, are seen about and around the composite'.[77]

Although Novak argues that '[c]omposite photography, then, has made biblical typology and kabalistic transmigration of souls at home within the technology of realism and the methodology of science', Galton would not have been at home with these irrational invasions into his carefully constructed representations of identity any more than he was comfortable with the 'ghostly accessories' he admitted kept turning up in his photographs.[78] Indeed, his composites seem to be hampered and haunted by these unwanted 'ghostly accessories', which prevent him from achieving 'a perfect test of truth'.[79] In *Generic Images* (1879) he is particularly concerned about the implications of the spectral accessories on his photographs. He suggests that '[i]f the number of combined portraits had been large, these ghostly accessories would have become too faint to be visible'.[80] While he wants to exorcize the ghosts from the images, he also implies that despite being invisible they will still be there, haunting the pictures. Galton is deeply anxious about how this haunting affects his results and attempts to argue that the ghosts do not alter the strength of his findings. He begins by suggesting that the ghosts 'are not sufficiently vivid to distract the attention', but Galton himself seems distracted by the ghosts in his own discussion.[81] He protests that regardless of the presence of the ghosts, the composites represent a 'truthfulness of which there can be no doubt', but his disavowals seem to point to the fact that it is he who doubts the success of the project because of the ghostly interferences.[82] Galton is haunted by the inability to secure identity in his composite portraits. The images themselves visibly resist capture, refusing to be fully submerged in a single image by fracturing off into spectral features.

Significantly, from 1872 until the early 1880s, Galton became interested in spiritualism, attended séances, and even worked with the SPR. There is no record as to whether he experimented with spirit photography or what he thought about

---

[75]   Ibid., p. 750.
[76]   Galton, *Generic*, p. 3.
[77]   Ibid., p. 3.
[78]   Novak, p. 103.
[79]   Galton, 'Composite', p. 140.
[80]   Galton, *Generic*, p. 3.
[81]   Ibid., p. 3.
[82]   Ibid.

it. Galton's ghostly composites, however, do bear striking similarities to the spirit photographs taken in the late Victorian period: translucent features and white, misty images appear in both types of photographs. Although Janet Oppenheim suggests that Galton's 'route did not again swerve from the clarity and precision of numbers into the mists and fogs of spirit', his composites, nevertheless, demonstrate that he was always in some measure drawn to the ghostly, and his initial inquiries into and later dissatisfaction with spiritualism suggest that he was both fascinated and repulsed by the supernatural.[83]

Galton wanted to guard against trickery in spiritualist practice, and during a séance held in January 1873, 'he wished to get rid of conditions which were manifestly fraudulent or favourable to fraud'.[84] His attempts to systematize this séance in order to see real ghosts, however, were apparently abortive: the spiritualist William Stainton Moses, who was present at the séance, wrote that it was Galton's hostile attitude which prevented any ghosts from materializing.[85] Significantly, Galton's attempts to use empirical methods during séances to authenticate ghosts ensured no ghosts would appear. Conversely, in his composites, his attempts to contain identity within an image caused spectral figures to emerge, suggesting that it is Galton's attempts to capture the immaterial that haunt him. The ghostly always seems to hover just around him, whether refusing to materialize at séances or stubbornly emerging in his composites. Both Doyle and Galton wanted to control when and how spirits would appear, but they are also representative of the ways in which scientists and psychical researchers were fascinated with and repelled by the supernatural at the fin de siècle. The spectre haunting late nineteenth-century thinkers was not simply the ghost in the séance room, but also the spectre of empirical science itself, which could neither prove the existence of the spiritual nor exorcise the spirits from scientific endeavour.

Criminal and Spiritual Influences on Holmes's Sensitive Plate

> He never spoke of the softer passions, save with a gibe and a sneer. They were admirable things for the observer – excellent for drawing the veil from men's motives and actions. But for the trained reasoner to admit such intrusions into his own delicate and finely adjusted temperament was to introduce a distracting factor which might throw a doubt upon all his mental results. Grit in a sensitive instrument, or a crack in one of his own high-power lenses, would not be more disturbing than a strong emotion in a nature such as his. And yet there was but one woman to him, and that woman was the late Irene Adler, of dubious and questionable memory.[86]

---

[83] Oppenheim, p. 296.
[84] Ibid., p. 294.
[85] Ibid.
[86] Doyle, 'Scandal in Bohemia', p. 161.

In 'A Scandal in Bohemia' (1891), Watson suggests that any 'intrusions' on Holmes's 'delicate and finely adjusted temperament [...] might throw a doubt upon all his mental results'. He goes on to imply, however, that Irene Adler, or '*the* woman', as Holmes refers to her, is one such intrusion – suggesting, in spite of his claims to the contrary, that Holmes *is* susceptible to irrational impulses like the 'softer passions'.[87] But if Holmes is receptive to illogical feelings like love, what other 'intrusions' could disturb his 'mental results'? Holmes's 'scientific mind' would seem to resist the notion of any subrational elements, particularly the supernatural.[88] In 'The Adventure of the Sussex Vampire' (1924), for example, he insists that '[t]his agency stands flat-footed upon the ground and there it must remain. The world is big enough for us. No ghosts need apply'.[89] Yet Holmes is also delighted by supernatural possibility and complains to Watson that 'in avoiding the sensational, I fear that you may have bordered on the trivial'.[90] Furthermore, many of the Holmes stories, like *The Hound of the Baskervilles*, 'The Adventure of the Sussex Vampire', and 'The Adventure of the Shoscombe Old Place' (1927), hint at the supernatural (a phantom hound, a vampire, and a haunted crypt all haunt but never materialize within the stories), suggesting that the Holmes stories are haunted by the very supernaturalism that Holmes's fastidious deductive techniques seem to preclude.

Holmes is, as Watson suggests, a 'sensitive instrument', and here I want the term 'sensitive' to resonate with its many different definitions. Not only does it signify that Holmes has 'quick or intense perception or sensation', but also that he is 'one sensitive to spiritualist or other occult influences, a medium' and that he has 'the temperament that is receptive of hypnotic or other occult influences'.[91] Furthermore, the term is linked to photographic language, particularly in relation to sensitive paper and the sensitive plate, and the ways in which the camera is sensitive to light.[92] Indeed, the 'sensitive instrument' with its 'high–power lenses' could refer both to Holmes and to a camera (significantly, 'A Scandal in Bohemia' centres on a photograph), suggesting that although both figures are linked to a logical means of representing the world, they are also sensitive to occult influence.[93] Like the camera, Holmes is 'delicate' and 'finely adjusted', but when

---

[87]   Ibid.

[88]   Doyle, *Hound*, p. 672.

[89]   Doyle, 'The Adventure of the Sussex Vampire', in *Complete Sherlock Holmes*, pp. 1033–44 (p. 1034).

[90]   Doyle, 'The Adventure of the Copper Beeches', in *Complete Sherlock Holmes*, pp. 316–32 (p. 317).

[91]   The notion of a sensitive as a medium dates from 1850 and its link to hypnotism dates from 1846 (OED).

[92]   Sensitive paper was in usage from 1839 and the idea that a camera was sensitive to light also dates from that year. The concept of sensitive plates dates from 1893 (OED).

[93]   In spirit photography, both the camera and the photographer needed to be 'sensitive'. According to Mrs. James Coates in *Photographing the Invisible* (London: L.N. Fowler and Co., 1911), a 'sensitive-photographer' is 'essential': 'that a medium is necessary is borne out by the fact that there are several thousand photographers in Great Britain alone, and

he allows 'intrusions' like criminal behaviour and the supernatural (in particular, spiritualism) onto his sensitive plate, they do not, as Watson fears, disturb his 'mental results', but rather give him uncanny powers of deduction.

The 'grit' that Watson worries will disturb Holmes's rational mind (or scratch the camera's lenses) might in fact refer to the seedy underbelly of the criminal world which influences Holmes. Holmes often participates in illicit activities – he takes drugs, he commits a felony in 'The Boscombe Valley Mystery' (1891), and he is even 'indirectly responsible for Dr. Grimesby Roylott's death', which he insists is not 'likely to weigh very heavily upon my conscience'.[94] Significantly, young Stamford, Watson's old acquaintance, remarks to Holmes, '[y]ou seem to be a walking calendar for crime'.[95] Rather than wishing to mend his ways, however, Holmes is inspired by proximity to and participation in criminal activity. He defends his drug use because his 'mind […] rebels against stagnation', something which the drugs prevent.[96] Furthermore, on naming a new case 'A Study in Scarlet', Holmes ponders, 'Why shouldn't we use a little art jargon. There's the scarlet thread of murder running through the colourless skein of life, and our duty is to unravel it, and isolate it, and expose every inch of it'.[97] Here he suggests that without crimes like murder, life is 'colourless' – bland and uneventful. While he is dedicated to stopping crime, it is the darker aspects of life which fascinate and propel him.

Holmes also uses photographic language to describe the way in which he unravels a trail of clues – he wants to 'expose' the crime, just as a photographer chooses the exposure time for an image.[98] While photography plays a role in many of the stories, often – as in the passage from *A Study in Scarlet* (1887) above – photographic language and contemporary ideas about the symbolic powers of the camera have been absorbed into Doyle's writing. When photographic terminology appears, it often does so in connection with Holmes's spiritual insight, suggesting that just as the camera is sensitive to spirits materializing in the séance, Holmes is also sensitive to supernatural influence. For example, Holmes's 'flame-like intuitions and impressions flash up', suggesting his link, not just to the camera (flash photography, the impression on the sensitive plate), but also to spiritualism.[99] Intuition connotes instinctive, rather than rational, reasoning, but also, and tellingly,

---

probably there are not five who are able to obtain these photographs' (p. 390). The work was influential on Doyle's own researches into spirit photography, and he recommends Coates's book to the reader in *The Vital Message*.

[94]   Doyle, 'Speckled Band', p. 273.

[95]   Doyle, *Study in Scarlet*, p. 18.

[96]   Doyle, *Sign of Four*, p. 89.

[97]   Doyle, *Study in Scarlet*, p. 36.

[98]   The terms expose and exposure were in use from 1839 (OED).

[99]   Doyle, 'The Adventure of the Creeping Man', *Complete Sherlock Holmes*, pp. 1070–83 (p. 1071). From the 1880s until the invention of the flash bulb in 1925, magnesium flash powder was used to take flash photographs. See Mary Warner Marien, *Photography: A Cultural History*, 3rd edn (New York: Harry N. Abrams, 2002), p. 207.

insight derived from spiritual origins.[100] In *The Medium and Daybreak*, a popular spiritualist weekly that ran from 1870 to 1895, intuition is defined as 'being always a voice from above' – it was a form of communicating with the spirit world.[101] According to Patricia Murphy, intuitive ability in the nineteenth century was particularly associated with women because it was believed they 'evidenced greater sensory ability, perception, and rapid thought'.[102] Amy Lehman argues that 'female mediums played a prominent part in the Spiritualist movement, as did "feminine qualities", such as intuition, sympathy, and sensitivity to non rational influences'.[103]

Watson is certainly aware of the link between femininity and intuition – in 'The Boscombe Valley Mystery' he describes how Miss Turner is able to recognize Holmes because she possesses 'a woman's quick intuition'.[104] He also recognizes Holmes's intuitive powers, but is eager to dismiss their potential connection to the spiritual, claiming that while he admires 'the rapid deductions, as swift as intuitions', these are 'always founded on a logical basis'.[105] Perhaps Watson's eagerness to associate Holmes with logical thinking speaks to his anxiety that this is the not the case: if Holmes's abilities are supernatural, then Watson's carefully constructed rational, masculine world might be in jeopardy. Watson argues that Holmes's 'brilliant reasoning power would rise to the level of intuition, until those who were unacquainted with his methods would look askance at him as on a man whose knowledge was not that of other mortals'.[106] But Holmes's knowledge is not that of other mortals. Instead, he can be connected to feminine, spiritual insight – he is a kind of medium, able to communicate with spirits.[107]

Significantly, Holmes's spiritual insight makes him a better detective, allowing him access to ideas and conclusions that would be closed to everyday detectives

---

[100] For a discussion of Victorian ideas about the role of intuition, particularly in the writings of Thomas Carlyle, see Susy Anger, *Victorian Interpretation* (Ithaca: Cornell University Press, 2005).

[101] William Oxley, 'W.J. Colville's Lectures', in *The Medium and Daybreak* (London: J. Burns, 1885), vol. xvi, 4 September 1885, p. 567. See also A. Victor Segno, *The Law of Mentalism: A Practical, Scientific Explanation of Thought or Mind* (Los Angeles: American Institute of Mentalism Publishers, c. 1902). Segno connects intuition to spiritualism, and defines it as a type of telepathy.

[102] Patricia Murphy, *In Science's Shadow: Literary Constructions of Late Victorian Women* (Columbia, Missouri: University of Missouri Press, 2006), p. 22.

[103] Amy Lehman, *Victorian Women and the Theatre of Trance: Mediums, Spiritualists and Mesmerists in Performance* (Jefferson: McFarland and Co., 2009), p. 84.

[104] Doyle, 'The Boscombe Valley Mystery', in *Complete Sherlock Holmes*, pp. 202–17 (p. 208).

[105] Doyle, 'Speckled Band', p. 258.

[106] Doyle, 'The Red-Headed League', in *Complete Sherlock Holmes*, pp. 176–90 (p. 185).

[107] In *Ms. Holmes of Baker Street: The Truth About Sherlock* (Alberta: University of Alberta Press, 2004), C. Alan Bradley and William A.S. Sarjeant argue that Holmes is actually a woman. I want to hint at Holmes's feminine traits to show how they link him to spiritualism and ideas about the spiritualist medium, who was usually female.

like Lestrade. When Watson says to Holmes, 'you have brought detection as near an exact science as it ever will be brought in this world', he implies that Holmes has been able to do so because of his connection to the other, or spirit world.[108] In response to Watson's praise, Holmes 'flushed up with pleasure at my words, and the earnest way in which I uttered them. I had already observed that he was as sensitive to flattery on the score of his art as any girl could be of her beauty'.[109] Holmes's sensitivity and femininity in this passage are, again, suggestive of the spiritualist medium, a role that sometimes falls on Watson, who, in the following passage, becomes the medium for Holmes's message. Holmes says, 'It may be that you [Watson] are not yourself luminous, but you are a conductor of light. Some people without possessing genius have a remarkable power of stimulating it'.[110] As a conductor of light, Watson might also be a spirit guide, the embodied form of a spirit. Furthermore, the notion of Holmes as 'luminous', or as the light which Watson conducts, is suggestive about the role of light in spiritualism. Many spiritualist works used the word to indicate a symbolic way of speaking about the spirit world, or to refer to the lights that appeared in séances.[111] As Oppenheim shows, it was also important how well lit a séance room was. While spiritualist investigators insisted there had to be enough light to ascertain the validity of spiritual events, 'the light in the séance room frequently had to be dim to enable spirit lights and luminous arms to emerge'.[112] Lighting was also crucial for spirit photography – there had to be enough (and the right kind) to get a clear picture, but not so much that it would discourage the ghost from materializing. As the light, Holmes becomes not only the spirit lights appearing at the séance, but also the link to the other world, the ghost speaking through Watson. Watson inspires Holmes to incredible feats of deduction, suggesting that Holmes is at his most brilliant when he accepts outside influence into his 'science of deduction'.

In 'The Crooked Man' (1893), Watson describes how Holmes's

> eyes kindled and a slight flush sprang into his thin cheeks. For an instant the veil had lifted upon his keen, intense nature, but for an instant only. When I glanced again his face had resumed that red-Indian composure which had made so many regard him as a machine rather than a man.[113]

---

[108] Doyle, *A Study in Scarlet*, pp. 33.

[109] Ibid., pp. 33–4.

[110] Doyle, *Hound*, p. 669.

[111] The spiritualist weekly paper *Light*, or to give its full title, *Light: A Journal Devoted to the Higher Interests of Humanity: Here and Hereafter*, was extremely popular and began publication in 1881 (Oppenheim, p. 46). Books with titles like *Light through the Crannies: Parables and Teachings from the Other Side* (London: Longmans, 1888), *The Bridge of Light: A Message from the Unseen* by Aster (London: Gay and Bird, 1899), and *Light from the Summerland* (London: Gay and Bird, 1901), by Lux Aurea, show how writers used the term to help define their way of thinking about spiritualism.

[112] Oppenheim, p. 14.

[113] Doyle, 'The Crooked Man', in *Complete Sherlock Holmes*, pp. 411–22 (p. 412).

The veil that lifted on Holmes's face evokes the spiritualist notion of a veil separating this world from the next, and which might be raised to reveal the spirit world. Here, Watson momentarily glimpses the spiritual, passionate aspect of Holmes, which is hidden beneath the cool objectivity of 'a machine'. As in 'A Scandal in Bohemia', 'the machine' might refer to the camera, and Holmes's composure suggests that he is composing himself for a picture. Although Watson exclaims in 'A Scandal in Bohemia' that Holmes is 'the most perfect reasoning and observing machine that the world has ever seen', he is also anxious about what this mechanization might be saying about Holmes.[114] When Watson tells Holmes, 'you really are an automaton – a calculating machine [...]. There is something positively inhuman in you at times', he suggests that Holmes's lack of personal feeling makes him robotic, perfectly, although mechanically, observing the world around him.[115] Holmes, however, insists that it is this very objectivity which makes him an excellent detective. But how does becoming a kind of 'automaton' alter Holmes's identity? While in 'Wireless' the mingling of minds, bodies, and machines has the uncanny effect of fragmenting identity, here it is Holmes's machine-like reasoning which is uncanny. In taking on the identity of an 'observing machine', Holmes becomes 'positively inhuman' – not only the relentlessly effective sleuth that Watson sees, but also, and significantly, an otherworldly and unearthly figure, his mind mechanically and furtively calculating, his identity slipping between man and machine. Paradoxically, his role as an 'observing machine', a camera, both condemns and invites association with the supernatural.

Holmes and Hypnotism

In 'The Illustrious Client' (1924), Holmes seems to recognize a fellow medium. He describes Violet de Merville as 'spiritual', 'a being of the beyond' possessing 'the ethereal other-world beauty of some fanatic whose thoughts are set on high'.[116] Her eyes are 'abstracted' as if she is in a spiritualist trance, but Holmes remembers that she might be under the influence of a 'post-hypnotic suggestion' induced by Baron Gruner.[117] While Holmes admits the possibility that 'One could really believe that she was living above the earth in some ecstatic dream', he remains unconvinced that the woman's actions are a result of the hypnotic trance, stating, '[y]et there was nothing indefinite in her replies'.[118] Surprisingly, this is not because Holmes's logical mind scoffs at the notion of hypnotism, but because he is a practiced hypnotist himself. In 'The Adventure of the Red Circle' (1911), Watson describes how 'Holmes leaned forward and laid his long, thin fingers upon the woman's

---

[114]   Doyle, 'Scandal in Bohemia', p. 23.

[115]   Doyle, *Sign of Four*, p. 96.

[116]   Doyle, 'The Adventure of the Illustrious Client', in *Complete Sherlock Holmes*, pp. 984–99 (p. 991).

[117]   Ibid., p. 992.

[118]   Ibid.

shoulder. He had an almost hypnotic power of soothing when he wished. The scared look faded from her eyes, and her agitated features smoothed into their usual commonplace. She sat down in the chair which he had indicated'.[119] Indeed, Holmes seems to fall into a hypnotic trance when he is at the height of his deductions, as if only by hypnotizing himself can he come to the correct conclusions. For example, Watson describes how his 'eyes had assumed the vacant, lack-lustre expression which shows mental abstraction'.[120] Holmes's 'mental abstraction' is not a sign of unfocussed thought processes, but rather indicates he is accessing heightened states of awareness. While I will return to a discussion of empowering trance states in women in ghost stories and New Woman texts later in this book, I want to suggest here that Holmes's access to hypnotic trance and to the spiritual grant him insights that are denied to Watson and police inspectors like Lestrade and are ultimately what make him such an excellent detective.

Doyle also wrote about trance states in his novella *The Parasite* (1894).[121] The narrator, Professor Gilroy, is both fascinated and horrified by the mesmerist Miss Penelosa, and while he believes that she has complete control over him, her death suggests that he was also a powerful force in the mesmeric relationship. I discuss the dynamic interchanges within mesmerism and hypnotism in the next chapter, but it is significant to note that in the Holmes stories Doyle chooses to focus on hypnotism, while in his Gothic novella he writes about mesmerism. A first reading suggests that this is a logical move: as I will detail in the next chapter, in the 1880s and 1890s, medical practitioners and mental scientists were attempting to present hypnotism as a scientific practice, while mesmerism was associated with the supernatural and sinister powers of mind. Nevertheless, hypnotism retained its links to supernatural power, both in the popular imagination and in the scientific community. Although Holmes's hypnotic abilities might on the surface link him to established, rational, scientific thought, they also connect him to all of the occult possibilities of mesmerism.

Significantly, in the quote from 'The Adventure of the Cardboard Box' which opens this chapter, Holmes's 'responsive' 'filaments' evoke not only ectoplasm and telepathy,[122] but also, as Oppenheim suggests, 'the summoning of invisible, imponderable fluids for explanatory purposes [which] likewise provided an important thread connecting spiritualism with mesmerism'.[123] Later in the story, Holmes says that he had been 'in rapport' with Watson when he was able to guess

---

[119] Doyle, 'The Adventure of the Red Circle', *Complete Sherlock Holmes*, pp. 901–13 (p. 902).

[120] Doyle, *Study in Scarlet*, p. 26.

[121] Doyle, 'The Parasite', in *The Edinburgh Stories of Arthur Conan Doyle* (Edinburgh: Edinburgh University Student Publications Board, 1981), pp. 41–80.

[122] Doyle, 'Cardboard Box', p. 888.

[123] Oppenheim, p. 218.

his thoughts, an expression which in mesmeric language suggests that the two were in sympathy with one another.[124]

Mesmerism and hypnotism were also linked with photography, and perhaps it is no coincidence that the supposedly hypnotic Baron Gruner takes and keeps snapshot photographs of the women he 'collects'.[125] Geoffrey Batchen writes that scientists made connections between electromagnetism and photography, and Alison Chapman suggests that 'the development of the "black art" of photography in Britain corresponds with the rise of mesmerism in the 1840s'.[126] According to Crary, the attentive or inattentive gaze into the camera and the abstracted gaze of the mesmerized or of distracted thought all become part of visualizing at the end of the nineteenth century.[127] In Holmes's case, his 'abstracted expression' not only means he is accessing an empowering trance state which will help him solve his case, but also that he is seeing what others do not – Holmes is capturing the invisible.

Holmes's relation to photography as it was conceived in the nineteenth century is as nuanced as the concept of photography itself. Doyle's relationship with photography was equally unstable, and his relationship with Holmes was always questionable; Doyle is well known for his ambivalence about his most popular creation.[128] In fact, Doyle was so uneasy about Holmes that he killed him off in 'The Adventure of the Final Problem' in 1893. However, he brought him back in 1901 for *The Hound of the Baskervilles*, and in 1903 began writing the stories for *The Strand* once more. That Doyle killed and resurrected his popular hero evokes the theory of the subject in the photograph who, as Barthes suggests, is at once going to die and is already dead.[129] Holmes is then always dead, and always dying, and always about to die: he haunts his own stories by the sheer fact of his being dead midway through them. The spiritualist in Doyle would have perhaps appreciated this turn of the screw. Or perhaps not: Doyle would have been not a little anxious about Holmes's spirit showing up in one of his photographs.

---

[124] Doyle, 'Cardboard Box', p. 889.

[125] Doyle, 'Illustrious Client', p. 990.

[126] Batchen, p. 154. Alison Chapman, '"A Poet Never Sees a Ghost": Photography and Trance in Tennyson's *Enoch Arden* and Julia Margaret Cameron's Photography', *Victorian Poetry*, 41 (2003), 47–71 (p. 50).

[127] Jonathan Crary, *Suspensions of Perception: Attention, Spectacle, and Modern Culture* (Cambridge, MA: MIT Press, 1999), pp. 11–76.

[128] See, for example, Daniel Stashower, *Teller of Tales: The Life of Arthur Conan Doyle* (London: Penguin, 2001).

[129] Roland Barthes, *Camera Lucida: Reflections on Photography*, trans. by Richard Howard (London: Flamingo, 1984). In Barthes the photograph is always a reminder of human mortality. In an examination of a photograph of Lewis Payne dating from 1865, Barthes argues, '*he is going to die*. I read at the same time: *This will be* and *this has been*; I observe an anterior future of which death is the stake. By giving me the absolute past of the pose (aorist), the photograph tells me death in the future. [...] I shudder, [...] over a catastrophe which has already occurred. Whether or not the subject is already dead, every photograph is this catastrophe' (p. 96).

Chapter 3
Identities and Powers in Flux: Mesmerism, Hypnotism, and George Du Maurier's *Trilby*

Introduction

> Then [Svengali] made little passes and counterpasses on [Trilby's] forehead and temples and down her cheek and neck. Soon her eyes closed and her face grew placid. After a while, a quarter of an hour, perhaps, he asked her if she suffered still.
>
> 'Oh! presque plus du tout, monsieur – c'est le ciel!' […]
>
> 'But never mind, matemoiselle; when your pain arrives, then shall you come once more to Svengali, and he shall take it away from you, and keep it himself for a soufenir of you when you are gone. […] *And you shall see nothing, hear nothing, think of nothing but Svengali, Svengali, Svengali!*'[1]

Just as Chapter 2 calls for a reassessment of assumptions about Sherlock Holmes, this chapter calls for a re-examination of conceptions about mesmerism and hypnotism. While it is generally acknowledged that hypnotism was born out of mesmerism and reached its 'heyday', according to Alan Gauld, in the 1880s and 1890s, what distinctions were actually made between the two practices by mental scientists, the medical community, and in fiction at the fin de siècle?[2] Where is the site of power in scenes of mesmerism and hypnotism? What was threatening about these practices and how did the literature reflect anxieties about them? In order to address these questions, it is first essential to give a brief summary of the history of mesmerism and hypnotism, both in scientific thought and in the popular imagination.

In the 1760s, Franz Anton Mesmer discovered a universal fluid which he believed could be harnessed by magnets and used for healing purposes. Using his hands or a magnetized wand, Mesmer would make passes over his subjects in order to restore

---

[1]   George Du Maurier, *Trilby* (London: Everyman, 1994), pp. 57–60. Subsequent references to this edition are given after quotations in the text.

[2]   Alan Gauld, *A History of Hypnotism* (Cambridge: Cambridge University Press, 1992), p. 575. See also Alison Winter, *Mesmerized: Powers of Mind in Victorian Britain* (Chicago: University of Chicago Press, 1998) and Betsy van Schlun, *Science and the Imagination: Mesmerism, Media and the Mind in Nineteenth-Century English and American Literature* (Berlin: Galda + Wilch Verlag, 2007) for discussions of how hypnotism originated out of mesmerism.

the natural flow of the universal fluid that disease or illness obstructed. Although Mesmer called the practice animal magnetism, the term mesmerism was adopted (at first derisively by those who scoffed at the notion that a universal fluid was part of the process). As Betsy van Schlun has pointed out, 'from the beginning, there are two strands inherent in mesmerism, a physical-scientific and a spiritual-mystic one': it was connected, for example, with theories of electricity and the workings of the brain, and with spiritualist séances and supernatural powers.[3] According to Athena Vrettos, mesmerism was '[o]ne of the most sensational Victorian fads' which 'reached its peak of popularity between the 1830s and 1860s, with a further revival of interest at the fin de siècle'.[4] Mesmerism should not, however, be understood as a fashionable craze that temporarily amused the Victorians, but rather as a crucial means of understanding Victorian ways of thinking. As Alison Winter has shown, mesmerism was not only 'practiced widely and continuously' (and by people of all classes and professions, including scientists like William Benjamin Carpenter and authors like Charles Dickens), it also played a 'pivotal role in transformations of medical and scientific authority'.[5] Although many theories circulated about how mesmerism worked, Winter argues that it was widely understood as a process in which the mesmerist (usually male) had the power to influence the thoughts and actions of the subject (often, although not always, female), using his hands to make 'passes' over her and sending her into a mesmeric trance.[6]

The passage from *Trilby* beginning this chapter not only describes this process, but also evokes many of the dangers Victorians associated with mesmerism. For example, Svengali's 'bold, brilliant black eyes' and heavy accent are suggestive about the notion of the mesmerist as a sinister and insinuating foreigner, possessed with magical power (p. 13). According to Daniel Pick, Svengali was a powerful symbol for the Victorian assumption that Jews possessed menacing powers of mental control which threatened Gentile, English identity.[7] The passage also indicates Victorian anxieties about the moral and sexual dangers of the male mesmerist, especially to vulnerable women: is it appropriate for Trilby to succumb to Svengali in such an intimate way? Furthermore, Du Maurier articulates the mysterious and supernatural possibilities of mesmerism: Svengali seems to instil in Trilby an unnatural impulse to '*see nothing, hear nothing, think of nothing but Svengali, Svengali, Svengali!*' (p. 60).[8]

---

[3]   van Schlun, p. 8.

[4]   Athena Vrettos, 'Victorian Psychology', in *A Companion to the Victorian Novel*, ed. by Patrick Brantlinger and William B. Thesing (Oxford: Blackwell, 2002), pp. 67–83 (p. 78).

[5]   Winter, p. 5.

[6]   Ibid., p. 3.

[7]   Daniel Pick, *Svengali's Web: The Alien Enchanter in Modern Culture* (London: Yale University Press, 2000), pp. 127–65.

[8]   See also Martin Willis and Catherine Wynne, eds, *Victorian Literary Mesmerism* (Amsterdam: Rodopi, 2006). This collection of essays focuses on many of the questions mesmerism raised, not only about issues like power, race, and sexuality, but also about class and gender.

Hypnotism was first developed by the Scottish surgeon James Braid, in part to debunk contemporary understandings of mesmerism. In *Neurypnology* (1843), he detailed the actions of the brain during the trance state, defining hypnotism as 'a peculiar condition of the nervous system induced by a fixed and abstracted attention of the mental and visual eye, on one object, not of an exciting nature'.[9] For Braid, the notion that a trance state could be induced by the mesmerist's will, or by invisible fluids, was both impossible and ridiculous. He hoped that his new term for the trance state, hypnotism, would not only remove supernatural and sinister associations from the practice, but also reveal that the trance state was psychological, induced with the cooperation of the hypnotized subject. According to Gauld, by the 1880s and 1890s 'hypnotism was a social and intellectual force of some significance', and influenced the discourse surrounding criminology and 'contemporary psychiatry and psychology'.[10] Indeed, during this period, mental scientists and medical practitioners, following Braid's early example, attempted to claim hypnotism for science. As Gauld has shown, '[p]sychotherapy as a self-conscious endeavour emerged from hypnotic practices in the later 1880s; indeed at first psychotherapy and hypnotherapy were synonyms'.[11] In the medical field, Jean Martin Charcot's experiments led him to argue that hypnotism was actually a manifestation of hysteria, and that only hysterical patients, who were usually women, were able to be hypnotized.[12] Ambrose Liébault and Hippolyte Bernheim of the Nancy School argued, however, that anyone could be hypnotized: hypnotism therefore revealed that susceptibility to trance states was part of the makeup of human identity.

Despite attempts to make hypnotism a scientific branch of study that was distinct from and in opposition to mesmerism, this chapter demonstrates how the discourse on hypnotism at the fin de siècle was actually a blend of ideas about mesmerism's supernatural powers and hypnotism's practical purpose in the field of psychology, developing in the 1880s and 1890s. While mental scientists attempted to police this blurring between superstition and practical study, close readings of hypnosis handbooks and mental scientists' researches on the subject show that the differences between the supposedly occulted mesmerism and scientific hypnotism

---

[9] James Braid, *Neurypnology: or the Rationale of Nervous Sleep Considered in Relation to Animal Magnetism or Mesmerism*, ed. by Arthur Edward Waite, new edn (London: Redway, 1899), p. 94.

[10] Gauld, p. 575. For further discussions of the ways in which mesmerism and hypnotism were influential, both in Freud's thinking and in 'psychological healing', see, for example, Adam Crabtree, *From Mesmer to Freud: Magnetic Sleep and the Roots of Psychological Healing* (New Haven: Yale University Press, 1993) and Maria M. Tatar, *Spellbound: Studies on Mesmerism and Literature* (Princeton: Princeton University Press, 1978). For a history of the rise of psychology as a discipline and profession, see Rick Rylance's *Victorian Psychology and British Culture, 1850–1880* (Oxford: Oxford University Press, 2000).

[11] Gauld, p. 575.

[12] Jean-Martin Charcot was a French neurologist who made significant contributions both to the study of hysteria and to the field of psychology.

were actually almost negligible. Mary Elizabeth Leighton has shown that 'medical men strove to appropriate hypnotism as the exclusive purview of their profession. Their efforts to establish their legitimacy to this claim, however, could not ultimately legitimize the figure of the hypnotist'.[13] While Leighton is interested in the ways in which the validity of hypnotists was undermined by their connection with criminal activity, both in the media and in fiction of the period, I suggest that whether or not hypnotism or its practitioners were legitimate is not the real question. Instead, the fact that mental scientists were unable to keep mesmerism and hypnotism distinct suggests that they were both fascinated and repulsed by the supernatural possibilities of trance states. Martin Willis argues that 'the mainstream scientific community attempted to construct a definition of science based on materialist criteria: empirical evidence, experimental verifiability, and inductive reasoning. This led, by the mid-nineteenth century, to the demise of occult science and the disaggregation of science from magical traditions'.[14] On the contrary, despite the materialist definition of science that emerged, mental scientists were still captivated, despite themselves, by the immaterial. Just as Francis Galton's composite photographs were haunted by unwanted, but nevertheless intriguing 'ghostly accessories', hypnotism was haunted and enthralled by mesmerism's supernatural account.[15]

The extract from *Trilby* beginning this chapter illustrates the overlapping between mental therapy and supernatural power within the discourse on hypnotism in the late nineteenth century. Svengali hypnotizes Trilby because she complains of the 'neuralgia in her eyes, a thing she was subject to ... the pain was maddening, and generally lasted twenty-four hours' (p. 55). While neuralgia is 'pain, typically stabbing or burning in the area served by a nerve', the disorder was also associated in the nineteenth century with hysteria (it is significant to note that Trilby's pain is described as 'maddening'). Daniel Hack Tuke's *A Dictionary of Psychological Medicine* (1892) defines neuralgia as 'derangement of sensibility',[16] and the entry on neuralgia, written by Heinrich Schüle, suggests that the symptoms can heighten during 'hysterical insanity'.[17] Furthermore, Schüle links neuralgia to gynaecological disorders, particularly hysteria, noting that the 'prolapse of the uterus' brings on 'attacks of mental derangement'.[18]

---

[13]   Mary Elizabeth Leighton, 'Under the Influence: Crime and Hypnotic Fictions of the *Fin de Siècle*', in Willis and Wynne, eds, *Victorian Literary Mesmerism*, pp. 203–26 (pp. 204–5).

[14]   Martin Willis, *Mesmerists, Monsters and Machines: Science Fiction and the Cultures of Science in the Nineteenth Century* (Kent, Ohio: Kent State University Press, 2006), p. 35.

[15]   Francis Galton, *Generic Images* (London: William Clowes and Son, 1879), p. 3.

[16]   Herbert Schüle, 'Neuralgia', in *A Dictionary of Psychological Medicine*, ed. by D. Hack Tuke, 2 vols (London: Churchill, 1892), II, pp. 835–40 (p. 835).

[17]   Ibid., p. 836.

[18]   Ibid., p. 839. Although by the end of the nineteenth century writings on hysteria focused more on the psychological impact of the disorder, hysteria was still connected to female sexuality. In his entry on hysteria in *A Dictionary of Psychological Medicine*,

Significantly, hysteria and other nervous disorders were treated in the 1890s with hypnotherapy, suggesting that Svengali employs hypnotism in the accepted therapeutic manner.[19] While this passage is suggestive about how hypnotism was practiced by the medical community in the late Victorian period, it also demonstrates how hypnotism was closely linked to the inexplicable nature of the mesmeric process. For example, when Trilby tells the Laird that Svengali hypnotized her, the Laird exclaims, 'I'd sooner have any pain than have it cured in that unnatural way, and by such a man as that!' (p. 60). Significantly, the Laird considers hypnotic therapy 'unnatural': for him, the medical practice was still inextricably connected to supernatural control and had only ambiguous curative value.

This chapter will explore not only the blurring of the boundaries between mental science and the supernatural in discussions of hypnotism, but also the blurring of sites of power in both mesmerism and hypnotism. The chapter calls for a radical re-examination of our understanding of the power dynamic in mesmerism and hypnotism as it was written about in the Victorian period. Critics like Pick, Winter, and Mary Elizabeth Leighton have not adequately questioned the assumption that the mesmerist or hypnotist is the powerful agent in scenes of mesmerism and hypnotism. A close reading of mesmerism and hypnosis manuals shows that, surprisingly, the mesmerist/hypnotist and the mesmerized/hypnotized share the site of power. The power flows between the two bodies (suggesting Mesmer's notion of a universal fluid) so that both are empowered by the trance state. While proponents of the practice agreed with the notion that hypnosis was not the result of one person's will being imposed on another's (as Braid suggested, the hypnotized subject had to be consensual in the process), they would have been horrified by the possibility that the hypnotist could in any way be affected.

This chapter also gives a close reading of *Trilby* to demonstrate the dynamic interchange of power between Trilby and Svengali during scenes of mesmerism and hypnotism, and to discuss how, in its popularity, *Trilby* mesmerized its audiences. While Phyllis Weliver also recognizes how the novel subverts expectations about the passive female role in mesmerism, she is concerned with 'ownership of Trilby's voice' and Trilby's self-fashioning in music, rather than in the complexities of and unease surrounding the mesmeric process.[20]

I discuss *Trilby* in terms of mesmerism and hypnotism, not only because Du Maurier does so, but also because his text addresses both mesmerism's supernatural associations and hypnotism's therapeutic associations as they were

---

for example, H.B. Donkin suggests that aside from abstinence, '[t]here are clearly other stresses which render women especially liable to hysteria. The periodical disturbance of menstruation, the times of pregnancy and parturition [...] contribut[e] to the number of sufferers' (I, pp. 618–27 [p. 620]).

[19] See 'Hypnotism in the Hysterical', *Dictionary for Psychological Medicine*, in which Charcot outlines some of the benefits of treating hysteria with hypnotism (I, pp. 606–10).

[20] Phyllis Weliver, *Women Musicians in Victorian Fiction, 1860–1900: Representations of Music, Science and Gender in the Leisured Home* (Aldershot: Ashgate, 2000), p. 247.

practiced at the fin de siècle.[21] While I want it to be clear that mental scientists tried (however unsuccessfully) to make distinctions between mesmerism and hypnotism, in the popular imagination and in fiction, particularly in the Gothic fiction of the 1880s and 1890s, the practices often merged into one another. Roger Luckhurst usefully refers to these Gothic works as 'trance-texts', and includes *Trilby*, Doyle's *The Parasite*, Richard Marsh's *The Beetle*, and Stoker's *Dracula* as numbering amongst them.[22] He argues that 'the late Victorian Gothic resonated with every nuance given to trance'.[23] This chapter suggests that the Gothic fear articulated in trance-texts was the dangerous merging of identities and powers in scenes of mesmerism and hypnotism. Although Pick argues that Svengali's mesmerism is representative of Victorian anxieties about the dangers of foreign influence, I suggest that it was not the notion of an outside force that was threatening, but rather the dangerous comingling of forces that crucially concerned Victorians in the 1880s and 1890s. If the site of power is shared in scenes of mesmerism and hypnotism, how does that affect notions of sexuality, nationality, race, class, and gender?

Mesmerism and Hypnotism: Blurring the Boundaries of Science and the Supernatural

Since Mesmer first wrote about mesmerism as a 'universal fluid' in the eighteenth century, the literature surrounding the practice has retained the link to invisible, immaterial action. For example, in *Facts in Mesmerism* (1840), which was one of the most widely read and influential of mesmerism texts, Chauncy Hare Townshend refers to mesmerism as an 'imponderable agent' which 'influences' the patient.[24] An 'influence' originally meant 'the supposed flowing or streaming from the stars or heavens of an ethereal fluid acting upon the destiny and character of men' and also an 'occult force'. Townshend uses the term to describe how a mesmerist could command this mysterious emission in order to regulate the actions of his subject.

Significantly, the psychiatrist Pierre Janet, whose work was crucial to late nineteenth-century mental science, writes in 1919 about hypnotism as if it is still Townshend's 'imponderable agent', invisible, and occulted: '[a]ttention is first drawn to a particular force by its exceptional manifestations. Not until then do people begin to acquire knowledge about the everyday phenomena that

---

[21]   For example, Svengali is described as having both 'hypnotic influence' (p. 75) and 'mesmeric powers' (p. 108).

[22]   Luckhurst, Roger, *The Invention of Telepathy: 1870–1901* (Oxford: Oxford University Press, 2002), p. 205.

[23]   Ibid.

[24]   Chauncy Hare Townshend, *Facts in Mesmerism: with Reasons for a Dispassionate Inquiry into it*, 2nd edn (London: Baillière, 1844), p. 281. Townshend was a well-known public figure: he was friends with Charles Dickens, and Edgar Allan Poe satirized his work in 'The Facts in the Case of M. Valdemar' (1845).

result from the working of this force'.[25] Although Janet suggests that hypnotism is capable of everyday action, his use of the term 'force' is so vague it seems that Janet himself is not sure what these everyday phenomena might be. Despite protestations that hypnotism was exorcised of mesmerism's 'pseudo-science', the discourse on hypnotism was still exploring ideas about ephemeral emanations and unknown powers. Indeed, from 1880 onward, scientists (such as F.W.H. Myers, Eugène de Rochas d'Aiglun, Albert Moll, and Julian Ochorowicz) were as vague about hypnotism as Janet.[26] They could come to no conclusion about finding an exact definition of hypnotism, nor could they come to a consensus about distinctions between hypnotism and mesmerism. For many scientists, removing the supernatural from hypnotism was a difficult, if not impossible, task.

By the end of the century, handbooks about hypnotism could not be distinguished from those on mesmerism because their vocabulary and goals were so similar. The aim of most of these works was to explain how mesmerism/hypnotism worked, to give evidence of curative value, and to encourage readers to learn, practice, and be treated with mesmerism/hypnotism. The similarity between the practices was such that James Coates, in his 1897 publication of *Human Magnetism or How to Hypnotise: A Practical Handbook for Students of Mesmerism*, remarks that '[p]ractically, hypnotism is mesmerism. The phenomena observed being similar, change of name cannot alter them'.[27] Indeed, the directions for how to induce either a mesmeric or a hypnotic trance are strikingly similar. John Barter's *How to Hypnotise: Including the Whole Art of Mesmerism* (1890), for example, although seeming to argue that mesmerism and hypnotism must not be confused, goes on to explain that passes and a concentrated gaze into the eyes of the subject are necessary for both mesmerism and hypnotism.[28] Although Barter insists mesmerism and hypnotism have some important differences, primarily that the mesmeric state can be induced on a child under three, whereas the hypnotic state cannot,[29] and that 'the mesmeric sleep is refreshing, but the hypnotic sleep begets weariness and lassitude',[30] he is unable to maintain the sharp distinctions he believes are present between mesmerism and hypnosis in his discussion of them.

---

[25] Pierre Janet, *Psychological Healing: A Historical and Clinical Study*, trans. by Eden and Cedar Paul, 2 vols (London: Allen & Unwin, 1925), I, p. 151. Janet studied under Charcot.

[26] Eugène de Rochas d'Aiglun was a French psychical investigator interested in hypnotism and psychical phenomena. Albert Moll was instrumental in introducing hypnotism into psychology and published his *Hypnosis* in 1889. Julian Ochorowicz was a Polish psychologist and investigator of hypnosis.

[27] Coates's manual was popular enough to be reprinted in 1904. See Coates, *Human Magnetism or How to Hypnotise: A Practical Handbook for Students of Mesmerism*, new rev. edn (London: Nichols, 1904), p. v.

[28] John Barter, *How to Hypnotise: Including the Whole Art of Mesmerism etc.* (London: Simpkin, 1890).

[29] Ibid., p. 11.

[30] Ibid., p. 12.

In the section 'How to Hypnotise in Public', Barter seems unable to decide which practice he is discussing, and while referring to mesmerism in one sentence, he uses hypnotism as its synonym in the next.[31] The interchange between the words happens again and again throughout the text, making the distinctions between the terms problematic to define.

The language in Barter's work makes the contemporary confusion within the medical community palpable, not only in his conflation of hypnotism and mesmerism, but also in the terminology he uses to describe them. The terms 'operator' and 'subject' are used by most mesmerism *and* hypnosis manuals published in the nineteenth century to describe the mesmerist/hypnotist and the mesmerized/hypnotized. Furthermore, in Barter's manual, discussions of the 'will' of the mesmeric operator that is able to govern the actions of the subject, and the 'suggestion' made by the hypnosis operator to influence the actions of the subject, imply that 'will' and 'suggestion' are just different terms for talking about the same practice.[32] As Coates suggests, for many writing on the mesmeric and hypnotic practices, '[p]ractically, hypnotism is mesmerism'.[33]

The fact that discourses on mesmerism and hypnotism often became interchangeable because of an inability on the part of researchers to clarify, even for themselves, what mesmerism/hypnotism really was suggests the uneasiness they felt with the practices and their inability to wholly separate these practices from the occult. One of the major concerns of late nineteenth-century works on hypnotism was the attempt to dissociate it from the sinister reputation mesmerism had gained in the popular imagination, and to invest hypnotism with more positive connotations: hypnotism was to be viewed as an effective treatment for poor health. And yet the continuous reference to objections many people have had to both practices, and the dubious reputation the practices have had within the community at large, only serve to underline that the writers and practitioners of mesmerism/hypnotism had to convince even themselves that the practices were free from the supernatural and mysterious.

Many of the manuals credit both mesmerism and hypnotism for achieving phenomena outside of the normal range of human powers. For example, Coates suggests that during a hypnotic trance state, 'higher phenomena' can occur, such as the transference of taste and smell from the operator to the subject, as well as thought-reading and telepathy.[34] However, he denies that clairvoyance is possible under hypnosis, a point on which he and Barter agree.[35] Barter explains, '[c]lairvoyance can only be attained by repeated *mesmeric* processes. Introvision may be attained by the subject by repeated *hypnotic* processes. Introvision means the faculty of seeing clearly and distinctly the various operations of nature taking

---

31 Ibid., pp. 16–17.

32 Ibid., p. 17.

33 Coates, *Human Magnetism*, p. v.

34 Ibid., p. 180.

35 Ibid., pp. 180–213.

place in the *interior* of another person or in the sensitive himself'.[36] Both Barter and Coates wish to separate activities like clairvoyance from discussions of hypnotism, but are unable to dissociate this practice from the supernatural entirely. Both admit that hypnotism can induce thought-reading and telepathy, but that only in mesmerism can clairvoyance be induced. This attempt to invest hypnosis with more scientific credibility than mesmerism is unsuccessful, however, since clairvoyance, the 'insight into things beyond the range of ordinary perception', is ultimately an interchangeable term for thought-reading: to see into someone else's mind is, after all, to have an insight beyond the range of ordinary perception. The fact that hypnotism was meant to be a more scientifically pure form of mesmerism was problematic, especially since writers like Barter and Coates were not only unable to distinguish hypnotism from mesmerism, they were also unable to explain away the supernatural manifestations occurring during hypnotism.

Medical texts about hypnotism also seemed to conflate the boundaries between the practical and inexplicable applications of hypnotism in psychology. Charcot's language in the description of the use of hypnotism on a hysteric, for example, illustrates this ambiguity:

> [w]hatever be the method employed for hypnotisation, even with the most predisposed persons, it may happen that sleep does not follow with all its characteristics at the first *séance*. In a number of cases there appear at first only vague phenomena, difficult to appreciate, which are to the true phases of hypnotism what the aura is to an attack of hysteria.[37]

Charcot's description is, as he acknowledges, vague, and seems to describe a spiritualist séance rather than a medical one. Indeed, the use of the term séance as a definition of a spiritualist gathering (first used in 1845) predates its use as a 'sitting' for medical treatment' (first used in 1875). Furthermore, according to Winter, most Victorians would have known about the mesmeric séance, or the process in which a mesmerist mesmerizes his subject: the language here is, again, more suggestive of mesmerism than of hypnotism.[38] The 'phenomena' Charcot describes could either be ghosts in the séance room or the symptoms of the onset of a hypnotic trance, and his definition is here haunted by hypnotism's connections to the supernatural.

That members of the SPR like Myers and Charles Richet (according to Gauld, 'the person with whom the modern hypnotic movement began') were also interested in hypnotism ensured that its link to the psychical, and not just the psychological, would remain strong.[39] One of the objectives of the SPR was '[t]he study of

---

[36]   Barter, p. 26.

[37]   Charcot, 'Hypnotism in the Hysterical', p. 607.

[38]   Winter, p. 6.

[39]   Gauld, p. 295. See also Charles Richet's *Thirty Years of Psychical Research: Being a Treatise on Metaphysics*, trans. by Stanley de Brath (London: Collins, 1923), where he discusses how somnambulism (a word for a deep mesmeric or hypnotic trance state) may in fact be linked to 'magnetic fluid' (p. 100).

hypnotism, and the forms of so-called mesmeric trance, with its alleged insensibility to pain; clairvoyance and other allied phenomena'.[40] Even as hypnotism was adopted into modern psychological practice by eminent practitioners like Charcot, it was still resonating with supernatural and sinister implications, ensuring that the mind at the fin de siècle was to be a supremely haunted site.[41]

Mesmerism Manuals and Hypnosis Handbooks

The site of power, like distinctions between science and the supernatural in mesmerism and hypnotism, is a point of uneasiness in the handbooks, particularly in cases of transference. Unlike Freud's use of the term, in which a patient transfers his or her unresolved emotions onto the analyst, transference in nineteenth-century mesmerism and hypnotism suggests that the operator can be affected and infected by the subject.[42] For example, Coates advises that mesmerists select the healthiest subjects, lest the mesmerist himself experience any of the sickness of the patient.[43] Furthermore, he recommends that the operator choose subjects 'most contrasted to himself in temperament', suggesting that if the subject and object are too alike, then the risk of their mesmerizing one another is even greater, especially if both are endowed with strong magnetic abilities.[44]

Even as late as 1980, in *A Handbook of Medical Hypnosis*, Gordon Ambrose and George Newbold discuss the possible dangers of the kind of transference that can occur in the hypnotic state:

> It does seem possible that a patient's symptoms may, during hypnotherapy, be transferred inadvertently from the patient to the one who is treating him. If the therapist is not fully in control of the situation [...] it seems just possible that a patient's own thoughts and feelings may, in certain circumstances, be implanted in the mind of the hypnotist. Although one cannot be dogmatic about this since there still remains a vast amount that is unknown about the workings of suggestion, this possibility is one that should not be lightly discounted.

---

[40] From 'Objects of the Society' (1882), in Sally Ledger and Roger Luckhurst, eds, *The Fin de Siècle: A Reader in Cultural History, c. 1880–1900* (Oxford: Oxford University Press, 2000), pp. 271–2 (p. 271).

[41] Freud also experimented with hypnotism, most notably in Josef Breuer and Sigmund Freud, *Studies on Hysteria*, ed. and trans. by James and Alix Strachey, *The Pelican Freud Library*, 15 vols (Harmondsworth: Penguin, 1974), vol. III. However, Freud found the practice was not always reliable, and he became anxious about using hypnotherapy.

[42] For information on Sigmund Freud and transference see Freud's 'The Dynamics of Transference', in *The Standard Edition of the Complete Psychological Works of Sigmund Freud*, ed. and trans. by James Strachey, 24 vols (London: Hogarth Press, 1953–1974), vol. XII, pp. 97–108.

[43] Coates, *Human Magnetism*, p. 135.

[44] Ibid.

Some years ago one of us [George Newbold] did meet a hypnotherapist who appeared to have succumbed to this particular hazard.[45]

The first edition of this work was published in 1956. That it had reached a fourth edition in the 1980s is suggestive of the hold hypnosis has had, even on contemporary culture, and also indicates that the anxieties we have about hypnosis today are similar to those held in the nineteenth century about both mesmerism and hypnosis. Indeed, transference suggests that the subject has the potential to harm the operator. In transference, the subject cannot be dismissed as a powerless, passive victim of mesmerism, but must be seen as a subject indeed, as much an active agent in the hypnotic/mesmeric process as the operator himself. Transference in discussions of mesmerism and hypnotism also brings up fin-de-siècle anxieties about the dangerous influence of not only other identities, but also disease. For example, the notion of shared fluids and infection in the discourse on mesmerism is perhaps hinting at late nineteenth-century concerns about the spread of venereal disease.[46]

Although transference might be dangerously infectious, in the handbooks it was also invested with curative power. In *How to Hypnotise*, John Barter gives an example in which 'transfer-hypnotic treatment' is used at a Paris hospital in order to cure a man, Mr. X, who is suffering from the neurological disorder St. Vitus's dance.[47] Transfer-hypnotic treatment occurs between three people: the operator, the patient, and a third party, the medium (or subject), who acts to connect the operator with the patient. In the case of the Paris hospital, the medium was a young French girl. She and Mr. X held hands, and she alone was hypnotized by the operator. As soon as she fell into the hypnotic trance she took on all the symptoms of Mr. X's illness and adopted his personality, even claiming that she was a 21-year-old man (Mr. X's age). When the operator woke her from the trance she resumed her normal state, and Mr. X was once again inflicted with the symptoms of the disorder, but to a lesser degree. After several 'transfer' sessions, Barter reports, Mr. X was cured of the disease.[48]

Although this example of 'transfer-hypnotic treatment' is meant to convince the reader of the power of the operator to cure even the most invasive neurological disorders, instead it emphasizes how the subject/medium is also invested with power in the hypnosis process. After all, it is the subject/medium who takes on the symptoms of the illness and eventually effects the cure of Mr. X. The site of

---

[45]   Gordon Ambrose and George Newbold, *A Handbook of Medical Hypnosis*, 4th edn (London: Baillière, 1980), p. 57.

[46]   For more on attitudes towards venereal disease in the Victorian period, see, for example, Judith R. Walkowitz, *Prostitution and Victorian Society: Women, Class, and the State* (Cambridge: Cambridge University Press, 1980).

[47]   Barter, p. 27. St. Vitus's dance is a neurological disorder associated with rheumatic fever that is characterized by jerky, involuntary movements affecting especially the shoulders, hips, and face.

[48]   Ibid., pp. 27–8.

power passes from the operator to the French girl and finally to Mr. X, and then back again, suggesting once more the flowing nature that Mesmer saw was a part of animal magnetism.

Indeed, the passage beginning this chapter, in which Trilby suffers from neuralgia, is not only an example of transference but also suggests that the site of power in mesmerism/hypnotism fluctuates between Svengali and Trilby. Significantly, Charcot defines hypnotism as a 'neurosis',[49] which, according to Tuke, is 'a functional disorder of the nervous system – that is to say a disorder such as migraine'.[50] A migraine might be another symptom of Trilby's neuralgia, since the pain can be localized around the eyes and last for long periods of time. Indeed, Trilby herself seems to be suffering from hypnotism, and by virtue of her neurosis is both a hypnotist and one who is hypnotized. Discussions of Trilby's nervous system also evoke the ways in which language in nineteenth-century psychology plays with the notions of the electricity of the mind and electricity in the discourse on hypnotism: the nerves of the nervous system send electric impulses to one another, galvanization is used as therapy for some sufferers of neuralgia, and in the mid-nineteenth century, electricity was used to explain some of the workings of mesmerism.[51] The notion of electricity is significant in a discussion of a fluctuating circuit of mesmerism, since both processes involve shared energies and networks between bodies and minds, evoking the shared energies between Svengali and Trilby.[52]

In attempting to cure Trilby's neuralgia, Svengali becomes infected by Trilby and begins to develop symptoms of the disorder. Significantly, when Svengali takes Trilby's pain as a 'soufenir', he implies this notion of infection (p. 60). A souvenir is a remembrance or memory, but also a 'slight trace of something', intimating that there are now traces of Trilby's neurological disease in

---

[49]   Charcot, I, p. 606.

[50]   D. Hack Tuke, 'Neuroses', *A Dictionary of Psychological Medicine*, II, p. 850.

[51]   Schüle, 'Neuralgia', II, p. 840. Electrobiology, for example, was introduced to America in the 1840s. In his introduction to James Braid's researches on hypnotism, *Neurypnology*, Arthur Edward Waite argues that 'electro-biology' produces the same effects as hypnotism (see 'Biographical Introduction', in Braid's *Neurypnology*, pp. 1–66 [p. 49]). Michael Faraday was a natural philosopher who argued that light and magnetism were connected, a phenomenon that became known as electromagnetism. Winter suggests that his researches were significant in 'the experimental study of mind and nervous sensibility' in the mid-nineteenth century (p. 278).

[52]   See Nicholas Ruddick, 'Life and Death by Electricity in 1890: The Transfiguration of William Kemmler', *Journal of American Culture*, 21 (1998), 79–87. Ruddick discusses the impact of electricity on Victorian America, but suggests that by the end of the century, the power of electricity was no longer astonishing to the popular imagination. He discusses J. Maclaren Cobban's mesmerism novel, *Master of His Fate* (1890), as an example of how electricity in the late nineteenth century had 'lost the metaphoric power to convey what Cobban's novel is really about, which is the flow of sexual energy' (p. 86). Ruddick does not take into account, however, the ways in which the language of mesmerism still heavily relied on the vocabulary of and theories about electricity.

Svengali's mind. Furthermore, 'trace' is a psychological term meaning 'a change in the brain as the result of some mental experience, the physical *after-effect* as such', suggesting that the hypnosis session has impressed itself deeply onto Svengali's psyche (indeed, a trace can also be defined as a 'mark or impression left on the face, the mind, etc.').

According to *A Dictionary of Psychological Medicine*, some of the symptoms of neuralgia are 'acts of violent resistance, and assaults', symptoms Svengali displays after hypnotizing Trilby: he becomes violent and abusive, pinning Little Billee's arms behind him and taunting him (p. 88).[53] Furthermore, 'neuralgia becomes the direct foundation, i.e., the cause, of delusions or fixed ideas'.[54] Shortly after hypnotizing Trilby, Svengali begins to suffer from both: he becomes obsessed with Trilby and sees visions of her skeleton (the narrator refers to them as 'vicious imaginations of Svengali's' [p. 105]). Although Svengali assures Trilby that she will '*see nothing, hear nothing, think of nothing but Svengali, Svengali, Svengali!*' (p. 60), it seems that both Svengali and Trilby become fascinated by one another: Trilby 'dreamed of him oftener than she dreamed of Taffy, the Laird, or even Little Billee!' (p. 105), and Svengali exclaims 'how beautiful you are! It drives me mad! I adore you!' (p. 104).

Indeed, other symptoms of neuralgia include both 'heartache' and a 'guilty conscience', which suggests both the sexual tension between Trilby and Svengali (he adores her, she dreams of him even more than of her lover, Little Billee) and fin-de-siècle anxieties about fluid circles of power.[55] Do Trilby and Svengali have a guilty conscience because they are complicating the boundaries between Western and Eastern values, between Christianity and Judaism? Is Trilby's guilt read onto her by the narrator for her nude modelling and for behaviour inappropriate for a woman in the nineteenth century? Is Trilby 'guilty' of transgressing gender roles, taking on Svengali's masculine traits?

Trilby's Influence

According to Coates, the mesmerist is powerless to do anything 'contrary to the [subject's] will', and ultimately mesmerism only highlights the subject's innate disposition.[56] Although Coates discusses intrinsic talents and the moral will of the patient, he still attributes some of the mesmeric and hypnotic phenomena to the supernatural. Coates analyzes *Trilby* in detail, giving Trilby herself as an example of how hidden talents can become polished under the hypnotic trance.[57]

---

[53]   Schüle, II, p. 838.

[54]   Ibid., p. 835.

[55]   Ibid., p. 836.

[56]   Coates, *Human Magnetism*, p. 209.

[57]   Ibid., p. 185.

For Coates, hypnosis brings out innate ability, but it also brings out the uncanny possibility that the human mind can possess supernatural power.

Coates gives the following account of *Trilby*'s hypnotic success:

> Du Maurier's *Trilby* has been denounced as an impossible creation, and *Svengali* an impossible operator; but for all that Du Maurier's novel is founded on one interesting fact in hypnotism, *i.e.* that *subjects do manifest in hypnosis certain powers of mind not suspected in normal life* [emphasis mine]. I am quite willing to grant the impossibility of a tone-deaf girl becoming a brilliant *diva*; but the fact remains that many subjects give extraordinary display of faculty in hypnosis, which neither they nor their most intimate friends imagined them to possess. The operator must ever remember that whatever powers are displayed in hypnosis these must be innate, for hypnosis [...] cannot create any faculty. Every human faculty, as well as those of sensation, can be stimulated or exalted in hypnosis. [...] I may say that *Trilby* had her prototype in Manchester about fifty years ago, and Dr. Braid was the clever, but in this instance reputable *Svengali*.[58]

Coates refers to an incident in Manchester in the late 1840s when the renowned singer Jenny Lind (1820–1887) was invited to the home of the 'father' of hypnotism, James Braid. Braid wanted Lind to give her opinion on one of his case studies, an illiterate factory girl who, while hypnotized, could sing like a virtuoso, but in her normal state was unable to carry out the same astonishing vocal feats. Together, Braid and Lind performed a series of experiments with the girl. Braid would hypnotize her and, using hypnotic suggestion, instruct her to mimic Lind. Lind would then perform a series of complicated vocal techniques, which the girl would afterwards imitate perfectly. Quoting from William Benjamin Carpenter's *Mesmerism and Spiritualism* (1877), Coates recounts Braid's description of the girl's singing: '[s]he caught the sounds so promptly, [...], and gave both words and music simultaneously and correctly, that several persons present could not discriminate whether there were two voices or only one'.[59]

The talents displayed by the girl are discussed as if her ability to mimic Lind is an uncanny power, her voice magically sounding as if it were composed of many voices, and suggesting the supernatural elements still resounding within the discourse of hypnotism. The fact that Braid is compared to Svengali emphasizes the possible occult and sinister powers of the hypnotist. Coates's comparison of Braid and Svengali takes all that is meant to be scientific, reputable, and beneficial within the discussion of hypnosis and reinvests it with all of the supernatural power, sexual threat, and, as Daniel Pick argues, the racial threat that hypnosis could also represent. Braid, the father of hypnotism, suddenly becomes comparable to the figure that, even in contemporary speech, is associated with manipulation and sinister power.

---

[58]   Ibid.

[59]   William Benjamin Carpenter, *Mesmerism, Spiritualism, etc.: Historically and Scientifically Considered: Being Two Lectures Delivered at the London Institution* (London: Longmans, Green, 1877), p. 86.

The event at Braid's home, and the comparison Coates makes between Braid and Svengali, signify once again the ambiguity of power relations in discussions of hypnotism. Braid's hypnotism of the girl could be viewed as a demonstration of his ability to coax out nascent talent, to become a kind of conductor who orchestrates the girl's singing, or, even more frighteningly, as her evil manipulator. However, the girl herself already has possession of her own voice; the talent is already hers. The relationship between Braid and the girl is not the stereotypical one we might assume, in which the hypnotist takes complete control of the passive girl. Rather, the relationship is one in which Braid and the girl must work together, the site of power belonging to both of them, and the connection between them one of a shared desire for beautiful singing.

Jenny Lind herself holds an important position in the discussion of mesmerism and hypnotism in the nineteenth century, as well as in a discussion of *Trilby*. Du Maurier even compared her to Trilby, and, as Pick shows, her life follows the same pattern as the heroine's: she was an illegitimate child who later became one of the most famous singers of the Victorian period. According to Pick, Lind seemed able to hypnotize her audiences, entrancing them with her voice and even reducing them to tears.[60] Merchandise bearing her name was sold at all of her concerts. However, Lind herself was seen by the public to have been hypnotized, not only by the company she kept (the composer Felix Mendelssohn was said to have captivated her) but also by her husband, Otto Goldschmidt, who many believed had hypnotized her into marrying him. Lind was not only thought to have hypnotic singing powers and to be under the influence of her lover, she also appeared in an opera about somnambulism, *La sonnambula* (1831) by Vincenzo Bellini. Lind's role in the public eye was that of both hypnotist and hypnotized: she had the ability to enthrall her audiences and, according to them, could be enthralled by her husband. At Braid's demonstration she even acted as a hypnotist, giving suggestions to the young girl as to what she should sing. Lind's dual role suggests that the hypnotic power is one of dynamic flux that passes between bodies.

Like Lind, Du Maurier's novel was enormously popular, and the mesmerized Trilby became mesmerizing in her popularity. L. Edward Purcell argues that the novel 'set the pace for the emerging bestseller publishing system in America'.[61] Just as Lind merchandise was a bestseller, 'Trilbyana' became all the rage: Trilby jewelry, hats, shoes, ice cream, and sausages, for example, were all sold.[62]

---

[60]   Pick, pp. 118–26.

[61]   L. Edward Purcell, 'Trilby and Trilby-Mania, The Beginning of the Bestseller System', *Journal of Popular Culture*, 11 (1977), 62–76 (p. 62).

[62]   For more on the contemporary view of the popularity of *Trilby*, see, for example, Jeanette Leonard Gilder, *Trilbyana: The Rise and Progress of a Popular Novel* (New York: The Critic, 1895). *Trilby* was transformed into several plays, for example *Trilby* (1895) by Paul M. Potter, and *Drilby* (1896), a parody of Du Maurier's work. Pick estimates that in the United States there were 24 versions of the play running by 1896 (p. 40). The book was also adapted for the screen, and films like *Trilby* (1914), *Svengali* (1931), and *Svengali* (1955) were produced.

What is crucial about a discussion of mesmerism, hypnotism, and fame is the fact that it is not the Svengalis (both literary and real) who captivate their audiences, but rather the Trilbys and the Jenny Linds. Trilby is often seen as the powerless victim in Svengali's wicked plans, yet when she enters into the imagination of the British and American public, she is the one whose reception entrances.

Indeed, a discussion of celebrity, mesmerism, and hypnotism in *Trilby* also brings up questions about the role of authorship. Where is the site of authorship once a novel becomes so fascinating to the public? Who is the real author, once the characters begin to take on lives of their own in the popular imagination? Pick intimates that Du Maurier was anxious about the popularity of his book: 'Du Maurier was in many ways a self-effacing man; he gained only meagre pleasure from his immodest riches, and found the cascade of attention deeply intrusive'.[63] Perhaps *Trilby* is concerned not only with trance states and sites of power, but also with the dangers of hypnotic and mesmeric fame and authorship to the private life: the unstable nature of both mesmerism and hypnotism implies that there can perhaps be no private life, and that instead, information is fluid and public, passing from body to body.

Blurring of Identities and Powers in *Trilby*

Gecko describes Svengali as 'a demon, a musician!', whose hold over Trilby transformed her into an unrivalled performer (p. 347). Gecko believed that this transformation was only possible because

> '*There were two Trilbys*. There was the Trilby you knew, who could not sing one single note in tune. She was an angel of paradise. [...] Well, that was Trilby, your Trilby! That was my Trilby too – and I loved her [...] But all at once [...] with one wave of his hand over her – with one look of his eye – with a word – Svengali could turn her into the other Trilby, *his* Trilby – and make her do whatever he liked ... you might have run a red-hot needle into her and she would not have felt it. ... (pp. 350–52).

Although Gecko's long speeches about Trilby suggest that she was deeply under the influence of Svengali's powers, they also serve to emphasize the problems with assigning Svengali as the instigator and manipulator of Trilby's personality and fame. Gecko's reference to 'two Trilbys' signifies that Svengali created and controlled another Trilby, unlike the one the men in the novel knew and loved, and that he manifested within her powers that, in her normal state, would have been impossible for her to demonstrate (p. 350). Of course, Gecko seems to be referring not only to nineteenth-century studies on the duality of the mind, but also to the two hypnotic states: the waking (or normal) state, and the trance (or hypnoid) state that supposedly Svengali would manipulate when

---

[63]   Pick, pp. 16–17.

hypnotizing Trilby. For Gecko, Trilby's secondary trance state is the product of Svengali's will and is divorced from Trilby's own volition. However, by the late nineteenth century, many researchers believed that actions and speech displayed during the trance state were products of the hypnotic subject's subliminal self.[64] In other words, the trance state revealed tendencies that had always been there, but that had been sublimated within the mind.

Indeed, in Trilby's case, it may be that the 'other Trilby' is the real one, and that hypnosis uncannily reveals her true identity, an identity which horrifies her English friends. The 'other Trilby' might really love Svengali, which is terrifying for the very British Little Billee, Taffy, and the Laird because of his racial otherness. Furthermore, the 'other Trilby' shows signs of having a sexual appetite for Svengali's advances, which shocks the English characters. Like Lucy in *Dracula*, Trilby is demonized because her sexual desires do not conform to the expectations for feminine behaviour at the end of the nineteenth century.[65]

If her dangerous sexuality comes forward through hypnosis, then so, too, does her singing ability. Gecko's claim that Svengali had transformed Trilby into a 'nightingale' (p. 351) becomes problematic when he himself admits that Trilby has always possessed an enchanting voice: '[s]he had not much ear. But she had such a voice as had never been heard. Svengali knew that. He had found it out long ago' (p. 348). It is not the case, then, that Svengali has made of her a songstress, but rather, as in Braid's experiments with the factory girl, that he has helped to perfect her existing ability. Furthermore, Gecko's belief that Svengali created the 'two Trilbys' (p. 350) is undermined by the fact that she was referred to as doubled long before Svengali became a strong presence in her life:

> Trilby speaking French and Trilby speaking English were two different beings. Trilby's English was more or less that of her father, a highly educated man [...]. Trilby's French was that of the Quartier Latin – droll, slangy, piquant, quaint, picturesque [...]. [I]t was difficult to decide which of her two incarnations was the more attractive. (p. 75)

Her duality is exemplified in her friendship with the three Englishmen, as she begins to divide her time between being a French model for the body and being a friend and all-around helping hand to Little Billee, Taffy, and the Laird. When she is with them, she takes up the study of English culture, reading 'Dickens, Thackeray, [and] Walter Scott', and when she is with her French friends, or modelling, she

---

[64] See Crabtree; Janet's chapter in *Psychological Healing* entitled 'Subliminal Tendencies' (pp. 256–60); and F.W.H. Myers, 'Multiplex Personality', in Jenny Bourne Taylor and Sally Shuttleworth, eds, *Embodied Selves: An Anthology of Psychological Texts, 1830–1890* (Oxford: Clarendon Press, 1998), pp. 132–8.

[65] Critics have discussed the ways in which Lucy's vampirism in *Dracula* is representative of male fear about female sexual desire. See, for example, Ashley Craig Lancaster, 'Demonizing the Emerging Woman: Misrepresented Morality in *Dracula* and *God's Little Acre*', *Journal of Dracula Studies*, 6 (2004), 27–33.

adopts the French culture she has always known: she jokes, uses slang, and earns her keep in a way that shocks and embarrasses her English friends (p. 75).[66]

Trilby's self is doubled by her adoption of two cultures and languages, not by Svengali's hypnotic passes. Her choice is probably also mitigated, however, by the fact that the Englishmen find her French identity less desirable and appropriate for a woman. Her French self takes on masculine traits, which she sublimates when the Englishmen are present in order to bring forward her more pleasant and feminine English self. Trilby does this, however, not because she must, but because she cannily knows how to adapt herself to situations. Just as Trilby is a powerful agent when she is mesmerized or hypnotized, she also cunningly reads the others, deciphering what they expect and desire from her.

Du Maurier's descriptions and illustrations of these characters demonstrate the ambiguity of sites of power within the text. When Trilby first appears in the studio, she is described as 'a very tall and fully-developed young female' whose 'eyes were too wide apart, the mouth too large, the chin too massive' (p. 14). Her enormous voice resounds from an enormous mouth that is compared to the largest edifices of architecture. Svengali tells her that 'the roof in your mouth is like the dome of the Pantheon; there is room in it for 'toutes les gloires de la France,' and a little to spare!' (p. 58). Furthermore, Jules Guinot refers to her as 'la grande Trilby', which, while meant to suggest a certain greatness of character, also implies a greatness of stature (p. 36). In the illustrations, Trilby towers over the other characters. Little Billee must gaze up at her as he asks for her hand in marriage, and in the illustrations of her performances, she towers over the band, Svengali, and her audience, seemingly double the size of everyone else (p. 37).[67]

Trilby's size complicates her supposed powerlessness when faced with Svengali, but also emphasizes her masculinity, just as in her French identity she exhibits masculine behaviours, like using coarse language. In *Trilby*, masculine and feminine characteristics are exchanged between Trilby and Little Billee, mirroring the fluid exchange of powers in mesmerism and hypnotism. For example, whereas Trilby 'would have made a singularly handsome boy' (p. 16), Little Billee possesses an 'almost girlish purity of mind' (p. 10). While she is tall and has a strong presence, he has 'delicate, regular features' and is 'graceful […]

---

[66] For more on cultural identity in *Trilby*, see, for example, Sarah Gracombe, 'Converting Trilby: Du Maurier on Englishness, Jewishness, and Culture', *Nineteenth-Century Literature*, 58 (2003), 175–208. Gracombe discusses how Svengali tries to convert Trilby to Jewishness, and Little Billee tries to convert her to Englishness. While I agree that the male characters do try to influence Trilby, ultimately she also gains power in the text.

[67] In *Woman and the Demon: The Life of a Victorian Myth* (Cambridge: Cambridge University Press, 1982), Nina Auerbach also discusses Trilby's large feet and stature. While Auerbach is interested in the mythic, magical, and regenerative powers of women in late nineteenth-century fiction, I argue that mesmerism itself needs to be re-examined in order to suggest that power relations within the mesmeric process are fluid, and that the relationship between Trilby and Svengali is one of an interchange of powers.

with very small hands and feet' (p. 7). Indeed, Little Billee and Trilby exchange conventional gender roles, so that Little Billee is portrayed as passive and feminine: he is inclined to burst into tears, and horrified when he discovers that Trilby models nude. Trilby, on the other hand, adopts the masculine role of independence and authority: she looks after herself and her brother, and forges her career as a professional singer.[68]

Trilby and Svengali's relationship is less a subversion of expectations of gender roles and more a reflection on the fluid circles of power passing from mesmerist to mesmerized in scenes of mermerism. Despite the fact that Trilby physically dominates the pages of the novel, the illustrations also show that a balance of influence exists between 'la grande Trilby' and Svengali. In one illustration Trilby and Svengali are depicted bowing after a performance. Here they are shown at level height with hands held, their postures indicating partnership and amicability. Furthermore, the boys collecting flowers behind them seem to represent innocence, stripping the image of Svengali and Trilby in the forefront of its possible manipulative connotations. This illustration suggests that Svengali and Trilby have collaborated, that they have worked together to achieve an excellent performance. The illustration is entitled 'And the remembrance of them – hand in hand', which is an excerpt from Little Billee's thoughts on the performance (p. 257): 'And the remembrance of them – hand in hand, master and pupil, husband and wife – smiling and bowing in the face of all that splendid tumult they had called forth and could quell' (p. 256). Significantly, Little Billee's statement evokes the mesmeric language itself – Trilby and Svengali's joint performance compels the audience to be enthralled. Even more significant is Little Billee's choice of the word 'them'. It is not Svengali alone who commands his audience, but both Svengali and Trilby, working together, who achieve 'that splendid tumult' (p. 256).

Indeed, they become increasingly intertwined as the novel progresses, so that not only do they form a partnership of conductor/singer, but they also absorb one another's identities. In becoming 'La Svengali', it seems that Trilby is subsumed by Svengali. However, a closer reading shows that, in fact, they have instead become subsumed by one another, their identities easily and increasingly interchangeable as the popularity of 'La Svengali' spreads. Svengali claims in the beginning of the novel that Trilby '*shall see nothing, hear nothing, think of nothing but Svengali, Svengali, Svengali!*' (p. 60), but when Little Billee hears her sing he believes it is as if she is saying, 'for I am *Trilby*; and you shall hear nothing, see nothing,

---

[68]   See also Dennis Denisoff, '"Men of My Own Sex": Genius, Sexuality, and George Du Maurier's Artists', in *Victorian Sexual Dissidence*, ed. by Richard Dellamora (Chicago: University of Chicago Press, 1999), pp. 147–69. Denisoff explores the blurring of gender and sexuality in *Trilby*, but ignores the implications of this blurring in relation to mesmerism. In 'Dazed and Abused: Gender and Mesmerism in Wilkie Collins', in Willis and Wynne, eds, *Victorian Literary Mesmerism*, pp. 163–82, Sharrona Pearl examines how mesmerism could 'challeng[e] existing norms' in gender relations (p. 163). However, while I show how mesmerism subverts expectations about gender in *Trilby*, Pearl suggests that Collins's work 'ultimately support[s] notions of female passivity and male activity' (p. 163).

think of nothing, but *Svengali, Svengali, Svengali!*' (p. 245). Here 'Svengali' is ambiguous, referring either to Trilby – La Svengali – or to Svengali himself. When the audience chants 'Svengali' over and over again during the performance in which Trilby cannot sing, Du Maurier writes that the crowd 'took up the cry [of Svengali], derisively' (p. 289). It is not clear, however, whether their derision is pointed at the now tuneless Trilby or at the master conductor who has failed to make the nightingale sing: the conductor and singer have become one identity in the eyes of the audience. The merging of names in this text is a merging of identities and power, which particularly concerns the conventional figures of Little Billee, Taffy, and the Laird, who are horrified by the possibility that the boundaries between identities might be permeable: the Christian Trilby can become Jewish, and her feminine characteristics can become masculine.

Trilby and Svengali bind themselves so closely to one another that by the end of the novel they are unable to function independently. When Trilby does not sing for Svengali at her final performance, Svengali dies. The power he exerts over her is also one which she wields over him, and once the balance of that power is broken, neither party can regain equilibrium. In other words, once the connection between them is severed, neither Svengali nor Trilby is able to withstand the fallout of power that ensues. Svengali dies with his eyes focussed on Trilby, still desiring her song, and Trilby dies shortly afterwards; her last words are 'Svengali ... Svengali ... Svengali ...' (p. 333). Both attempt to conjure one another as they die, trying to evoke the magnetic current that binds and enthralls them.

Indeed, much of the mesmerism and hypnotism fiction at the end of the century is concerned with the theme of the subject and operator becoming dependent on one another. For example, in J. Maclaren Cobban's *Master of His Fate* (1890), the magnetic Courtney is physically and emotionally drained by his love for Nora. Marie Corelli's *Ziska: The Problem of a Wicked Soul* (1897) is about a mesmeric love affair in which the lovers must attain a balance of power in order to achieve happiness. Henry James's *The Bostonians* (1886) complicates the site of power in mesmerism and hypnotism by portraying Verena Tarrant as both an entrancing public speaker and a passive participant in her private life.

Recognizing the shared site of power in mesmerism and hypnotism seems to offer a harmonious reading of the Victorian trance state. Alison Winter suggests that despite all the disagreement surrounding understandings of mesmerism, 'mesmerism and similar cultural phenomena are part of a history of agreement. They displayed a cord that bound people together: an influence that coordinated their thoughts or actions, or a sympathetic current that united a population'.[69] But by the 1880s and 1890s, works like *Trilby*, *Dracula*, *The Beetle*, and *The Parasite* were surging with the Gothic horror of the trance, not with friendly exchanges of power. Indeed, shared sites of power were sites of Gothic anxiety: if everyone was influenced by everyone else, then identity itself was at risk, not only of mingling

[69]   Winter, p. 304.

dangerously with other sexualities, nationalities, races, classes, and genders, but also of deteriorating or dispersing altogether. Distinctions between identities might become negligible, and the sense of self could be lost – something that Little Billee, Taffy, and the Laird deeply feared would happen to Trilby. The fascination that trance states held for writers in the 1880s and 1890s suggests that they were trying to safely contain the dangers of mesmerism and hypnotism within the fixity of print, while simultaneously delighting in the creative possibilities trance states offered. Furthermore, mental scientists hoped to control hypnotism and studies of the mind, but were unable to remove the supernatural emanations of mesmerism from their work.

Chapter 4 is also concerned with the ways in which late nineteenth-century mental science was attempting to maintain rigid boundaries in the study of the mind. Mental scientists wanted to study both spiritualism and the mind empirically, but succeeded only in ghosting the mind, and, as I will argue, particularly the *female* mind and body. While this chapter is concerned with the ways in which trance states like hypnotism change sites of power, the next chapter suggests that women see ghosts in altered states of perception: women reclaim ghostliness as an empowering female trait.

Chapter 4
Ghostwomen, Ghostwriting

Introduction: W.T. Stead and the Woman That Haunts Us

> This is the question that lies at the root of all the controversy as to ghosts. Before disputing about whether or not there are ghosts outside of us, let us face the preliminary question, whether we have not each of us a veritable ghost within our own skin?[1]

William Thomas Stead, the publisher and journalist most famous for his exposé of child prostitution in *The Maiden Tribute of Modern Babylon* (1885), was fascinated by the possibilities of the ghosts within us and the concept that the subliminal consciousness haunted the waking one. His interest in promoting knowledge about 'the other world' led him to publish the spiritualist journal *Borderland* in 1893. Even after perishing on board the *Titanic*, Stead seemed to continue his spiritual mission: his spirit supposedly contacted spirit photographer William Walker by psychically imprinting a written message on a photograph. Doyle included the photograph in his *Spirit Photography* with the message (which handwriting experts confirmed was in Stead's hand) 'Dear Mr Walker, I will try to keep you posted'.[2]

Stead's *Real Ghost Stories* (1891) was enormously popular, selling 100,000 copies in a week.[3] It included hundreds of accounts of encounters with ghosts and was rigorously categorized into subsections, for example, 'Ghosts of the Living on Business', 'Ghosts that Keep Promises', 'Apparitions at or About Death', 'Ghosts of the Dead or Ghosts Announcing Their Own Death', 'Ghosts of the Dead with a Practical Object', 'Ghosts in the Open Air or Out of Door Ghosts', and 'Tangible Ghosts or Ghosts which Touch' (p. i). The publication even included a 'Census of Hallucinations', which readers were meant to complete to provide Stead with statistics for the number of people subject to ghost-seeing and information about the nature of their visions. Stead guaranteed that all the stories were absolutely true, and because of this he warned weak-hearted readers to put the publication down immediately and forever.

---

[1]   W.T. Stead, 'The Ghost that Dwells in Each of Us', in *Real Ghost Stories: A Record of Authentic Apparitions* (London, 1891), pp. 11–21 (p. 11). Subsequent references to this edition are given after quotations in the text.

[2]   Arthur Conan Doyle, *The Case for Spirit Photography* (London: Hutchinson, 1923), p. 77.

[3]   Diana Basham, *The Trial of Woman: Feminism and the Occult Sciences in Victorian Literature and Society* (London: MacMillan, 1992), p. 154.

Stead's insistence that all of the ghost stories were authentic, factual accounts is akin to Doyle's persistent desire to discover the factual evidence that could prove spiritual existence. Furthermore, Stead's discussion of the ghost within reflects a newly occulted understanding of mental science which had been developed in the 1880s and 1890s by psychical researchers, psychologists, and physicians seeking to map the mind. Medical practitioners and psychical researchers like William James and Charles Richet sought to explain the spirit world by situating it in the realm of the psychological. Edmund Gurney, F.W.H. Myers, and Frank Podmore's interest in consciousness led them to elaborate a theory that 'ghosts' were projections of the mind, the result of thought transference or telepathy. In 1886 they published *Phantasms of the Living*, in which they

> propose [...] to deal with all classes of cases in which there is reason to suppose that the mind of one human being has affected the mind of another, without speech uttered, or word written, or sign made; – has affected it, that is to say, by other means than through the recognised channels of sense. [...] we have included among telepathic phenomena a vast class of cases [...]. [We] refer to *apparitions*; excluding apparitions of the *dead*, but including the apparitions of all persons who are still living, as we know life, though they may be on the very brink and border of physical dissolution. And these apparitions [...] include[e] not visual phenomena alone, but auditory, tactile, or even purely ideational and emotional impressions. All these we have included under the term *phantasm*.[4]

The use of the word 'phantasm', now employed psychoanalytically to denote psychic tension, indicates the effect supernatural studies had on the development of mental science. Gurney, Myers, and Podmore here compare telepathy with apparitions, suggesting that ghosts were the products of this psychic tension or phantasm: ghosts could be defined as not only external but also internal manifestations of the workings of the brain.[5]

Like Gurney, Myers, and Podmore, Stead also cites cases where apparitions of the living appear.[6] He recounts the case of a man so anxious about oversleeping and being late for work that, as he was still sleeping, his spirit materialized at work in his place (p. 32). In Mary Louisa Molesworth's 'Witnessed by Two' (1888), there is a fictional example of the living phantasms which captured the fin-de-siècle imagination. At first, it seems that the ghost of the heroine's lover appears to her at the moment of his death. Subsequently, the heroine discovers that a man sharing her

---

[4]   Gurney, Edmund, F.W.H. Myers, and Frank Podmore, *Phantasms of the Living*, 2 vols (London: Trübner, 1886), I, xxxv.

[5]   While Srdjan Smajic explores the ways in which ghost stories complicated the notion of visual stability in the nineteenth century, I am interested in how late-Victorian mental science was increasingly haunted by the ghostly.

[6]   Diana Basham reports that Stead had not yet read *Phantasms of the Living* at the time of his publication of *Real Ghost Stories* (Basham, p. 154). The fact that the works shared ideas suggests that the notion of a haunted mind was understood in the popular imagination, in psychical research, and in emerging psychology in the 1880s and 1890s.

lover's name has died. The hero's 'ghost' appears as a result of his extreme anxiety when he realizes that his lover will believe it is he, and not the other man, who is dead.[7] Both the living and the dead are engaged in haunting, and this haunting is directly connected to theories on the workings of the mind at the fin de siècle.

In the chapter entitled 'The Ghost that Dwells in Each of Us' in *Real Ghost Stories*, Stead asks 'whether we have not each of us a veritable ghost within our own skin?':

> Thrilling as are some of the stories of the apparitions of the living and the dead, they are less sensational than the suggestion recently made by hypnotists and psychical researchers of England and France, that each of us has a ghost inside him. They say that we are all haunted by a Spiritual Presence, of whose existence we are only fitfully and sometimes never conscious, but which nevertheless inhabits the innermost recesses of our personality. The theory of these researchers is that besides the body and the mind, meaning by the mind the conscious personality, there is also within our material frame the soul or unconscious personality, the nature of which is shrouded in unfathomable mystery. The latest word of advanced science has thus landed us back to the apostolic assertion that man is composed of body, soul, and spirit; and there are some who see in the scientific doctrine of the unconscious personality a welcome confirmation from an unexpected quarter of the existence of the soul. (p. 11)

Here, Stead not only articulates the haunted science of mind of the late Victorian period (referring, it seems, to Myers's notion of the 'secondary self' within the unconscious), but also, in placing it alongside accounts of interactions with ghosts, suggests that Victorians were fascinated by hauntings in all aspects of their lives: ghosts were everywhere invading ways of Victorian thinking, not only about conscious and unconscious states of mind, but also about conceptions of traditional societal roles. Significantly, in *Real Ghost Stories*, Stead argues that our normal waking self is male, but that our secondary self is female, comparing the unconscious mind with theories about hypnotism and gender relations in the following terms:

> The new theory supposes that there are inside each of us not one personality, but two, and that these two correspond to the husband and wife. There is the Conscious Personality, which stands for the husband. It is vigorous, alert, active, positive, monopolising all the means of communication and production. So intense is its consciousness that it ignores the very existence of its partner, excepting as a mere appendage and convenience to itself. Then there is the Unconscious Personality, which corresponds to the wife who keeps cupboard and store-house, and the old stocking which treasures up the accumulated wealth of impressions acquired by the Conscious Personality, but who is never able to assert any right to anything, or to the use of sense or limb except when her lord and master is asleep or entranced. [...] It is extraordinary how close this analogy

---

[7]   Mary Louisa Molesworth, 'Witnessed by Two', in *Four Ghost Stories* (London: Macmillan, 1888), pp. 43–86.

is when we come to work it out. The impressions stored up by the Conscious
Personality and entrusted to the care of the Unconscious are often, much to our
disgust, not forthcoming when wanted. It is as if we had given a memorandum
to our wife and we could not discover where she had put it. But night comes,
our Conscious Self sleeps, our Unconscious housewife wakes and turning over
her stores produces the missing impression; and when our other self wakes it
finds the mislaid memorandum [...]. [I]t is only when the Conscious Personality
is thrown into a state of hypnotic trance that the Unconscious Personality is
emancipated from the marital despotism of her partner. Then, for the first time
she is allowed to help herself to the faculties and senses usually monopolised by
the Conscious Self. (p. 20)

When Stead compares the unconscious personality with women, he also
addresses women and their social and political invisibility. Stead suggests that not
only is the mind a haunted site, but the ghost who haunts it is a woman, projecting
Victorian concepts about female identity and her passive role in marriage.[8] He also
notes, however, that while the woman ghosting the mind may be powerless for
much of the time, she can also powerfully disrupt the masculinist consciousness: the
female ghost here is invisible but noticeable, a surprisingly formidable and influential
presence. This chapter explores, through a case study of the SPR investigation into
the haunting of Ballechin House in 1897, the notion that for many Victorians the
haunted aspects of the mind were comparable to late Victorian ideas about gender
roles, and in particular the 'ghostly' role of women in society. Furthermore, a
discussion of contemporary theories of hysteria and spiritualist trance states suggests
that attempts by medical practitioners and mental scientists to police the female
mind and body seemed to make female identity more elusive and apparitional. At
the same time, however, the ghost was an empowering symbol for spiritualist and
writing women, allowing them access to greater freedom of expression.

If we can discuss a metaphorical female haunting of the mind and body at the
fin de siècle, then we must take into account what the implications are for women
writing about haunting at the end of the century. What do their ghost stories reveal
about their identity and ideas about women? This chapter explores the effect of
the occulted mental science developing in the late Victorian period on women's
ghost stories and women's self-identification (whether conscious or subconscious)
with the ghostly. Although women's ghost stories often articulate, as Vanessa D.
Dickerson suggests, 'eruptions of female libidinal energy [...], thwarted ambitions
[and] cramped egos', these are both a product of frustration with women's social
role and also, what she and other critics have so far ignored, an implication that

---

    8   Basham notes that Stead 'was quick to perceive the gender-implications of such
studies [psychic research and, in particular, hypnotism] and to use them as the basis for
a new social synthesis in which the emancipation of women and the liberation of the
unconscious personality were to play a significant part' (p. 154). Basham overlooks,
however, the impact of mental science on the ways in which women were perceived at the
fin de siècle. Furthermore, she focuses on masculine narrators in women's ghost stories,
ignoring the crucial aspect of how women were portraying women.

women were haunted by themselves.[9] A discussion of non-canonical women ghost-story writers and their work shows how authorship, ghostliness, and female identity are closely intertwined.

While according to Stead's analogy hypnosis allows the unconscious mind to awaken and take control (symbolizing women's emancipation from the 'despotism' of marriage), this also suggests subversively that trance states empower women. Significantly, hypnosis does not emphasize a woman's passive role in the home but actually heightens awareness about the 'despotism' in her married life. Indeed, as this chapter shows, altered states of perception like hypnosis, dreams, hysteria, and ghost-seeing become catalysts for creative expression and for political awakening in women's writing. In the final section of this chapter, using Kristeva's theory of abjection, I argue that the 'powers of horror' that women ghost-story writers experience and evoke create a female identity at the fin de siècle that is powerful, empowering, and hopeful as much as it is 'horrific'.

The Case of Miss Freer and the Haunting of Ballechin House

The example of Miss Freer and the Ballechin House hauntings not only demonstrates spiritualists' and scientists' interest in studying the supernatural scientifically and situating the ghostly in the mind, but also connects theories about the haunted aspects of the mind and the ways in which women were perceived at the end of the nineteenth century. In 1897 the SPR decided to investigate reports that Ballechin House in Perthshire, Scotland, was haunted. F.W.H. Myers recommended that Miss Freer (1865–1931), a member of the SPR since 1888, should stay in the house to record the phenomena.[10] She arrived in early February of 1897 and left in mid-April of the same year. Miss Freer made a detailed account of her stay, which was published (under the name of A. Goodrich-Freer), along with the accounts of other guests in the house, in *The Alleged Haunting of B— House* in 1899. The book minutely lists all visual, auditory, and tactile phenomena recorded at Ballechin, and even includes a floor plan of the house so that the reader can identify where each manifestation took place.

Although *Alleged Haunting* attempts to contain the hauntings at Ballechin within the fixity of print and as a series of facts (Miss Freer suggests that '[t]he editors [Miss Freer and John Patrick Crichton-Stuart, third Marquess of Bute] offer no conclusions. This volume has been put together, as the house at B— was taken, not for the establishment of theories, but for the record of facts'),

---

[9]   Vanessa D. Dickerson, *Victorian Ghosts in the Noontide: Women Writers and the Supernatural* (Columbia: University of Missouri Press, 1996), p. 8.

[10]   Freer reports in *The Alleged Haunting* of *B— House, Including a Journal Kept During the Tenancy of Colonel Lemesurier Taylor* (London: Redway, 1899) that 'Mr. Myers [...] wrote urgently' to her, 'saying, "If you don't get phenomena, probably no one will"' (p. 82). Miss Freer was said to possess great powers in crystal-gazing, clairvoyance, and automatic writing, and on the strength of these credentials W.T. Stead employed her as the assistant editor of the spiritualist journal *Borderland* from 1893 to 1897.

the controversy surrounding this investigation, and Miss Freer herself, was deeply concerned not just with finding facts, but also with regulating which facts would be shared with the public.[11]

The SPR's activities at Ballechin provoked the public to debate in *The Times* during June of 1897: contributors were concerned about the methodology of the tests, and a guest at the house during Miss Freer's stay, Mr. J. Callendar Ross, denounced the whole investigation as fraudulent.[12] Instead of defending Miss Freer, the SPR, and particularly Myers, turned their backs on her. Myers, who had earlier been enthusiastic about the findings and had written letters suggesting he believed that there was evidence of supernatural phenomena, now wrote to *The Times* on 10 June 1897, 'I visited B—, representing that society [SPR] and decided that there was no evidence as could justify us in giving the results of the inquiry a place in our *Proceedings*'.[13] Furthermore, he asked that his earlier letters be suppressed: 'I am afraid that I must ask that my B— letters be in no way used. I greatly doubt whether there was anything supernormal'.[14] Not only did the SPR attempt to use rigorous scientific methods to prove the existence of the supernatural, but they also wished to erase experiments which showed their methods in an unfavourable light: the SPR was trying to capture ghosts and fabricate science. For example, John L. Campbell and Trevor H. Hall recount how the SPR removed material from a volume of the *Proceedings* and issued a reprint containing a different case study to cover for the exposure of the 'nonsense' in the original volume.[15]

The SPR's actions in the Ballechin case particularly affected Miss Freer and made her a ghostly figure in the history of the SPR. Indeed, although Campbell and Hall's study, *Strange Things: The Story of Fr Allan McDonald, Ada Goodrich Freer, and the Society for Psychical Research's Enquiry into Highland Second Sight*, gives the most comprehensive information available on Freer, they write about her as if they want to exorcise her from both the history of the society and their own text. They are deeply hostile towards her, ostensibly because she copied Fr Allan McDonald's notebooks on Highland second sight with the intention of taking sole responsibility for authorship when she published them. Indeed, she seemed to believe that in consulting with McDonald on the subject, she was ghostwriting the project anyway: for Freer, authorship was a collaborative enterprise, and credit could never be assigned to a single individual.

Campbell and Hall vilify her throughout their book, not only because of the accusations of plagiarism, but also because of her feminine charms: according to

---

[11]   Ibid., p. 235.

[12]   See John L. Campbell and Trevor H. Hall, *Strange Things: The Story of Fr Allan McDonald, Ada Goodrich Freer, and the Society for Psychical Research's Enquiry into Highland Second Sight* (London: Routledge, 1968), pp. 185–96. Mr. Ross argued that Ballechin had no reputation for being haunted until the SPR arrived.

[13]   Goodrich-Freer, *Alleged*, p. 195.

[14]   Ibid., p. 193.

[15]   Ibid., p. 127.

the authors, she possessed 'personal attractions which seem to have been almost hypnotic in their effect and which she used irresistibly and ruthlessly upon those who she thought could be of use to her'.[16] Freer was threatening, both because she might taint the reputation of the SPR and because she was a woman with influence in a society formed and dominated by male membership: Campbell and Hall are particularly anxious about this threat, even hinting that she and Myers may have had an affair.[17]

What is significant in this case is that Miss Freer is an increasingly spectral presence in the SPR after the Ballechin incident. The records of her life are themselves vague: she called herself Miss Freer, Miss X., Mrs. Ada Goodrich-Freer, Mrs. Ada M. Goodrich-Freer, and Mrs. Ada Goodrich-Freer Spoer (or occasionally Spoor), suggesting that even her name was an ephemeral and fleeting form of identity. Campbell and Hall report that no one ever really knew her real age or anything about her family lineage.[18] She described her own family in the following terms: 'I belong to no effete race, but to a family which for physique and longevity might challenge any in the annals of Mr Francis Galton'.[19] Here it seems she might be alluding to Galton's composite portraits with their 'ghostly accessories', indicating that her family history is just as vague and elusive.[20]

In fact, a photograph is all that remains of Miss Freer's involvement in the SPR: '[t]he present officers of the Society say that they do not now possess any documents relating to her apart from the single photograph'.[21] Indeed, her presence in the Society fades out entirely with the turn of the century. In the same year that she published *Alleged Haunting*, she also published *Essays in Psychical Research*, and in 1900 she edited *The Professional: and Other Psychic Stories*, a collection of ghost stories, three of which she wrote herself.[22] After around 1902, however, she ceased to be actively involved in the Society and turned her attention to travel writing and other literary pursuits.

Ballechin house seems most haunted by Miss Freer herself. SPR member Frank Podmore said that 'during the investigation it was Miss Freer who first heard the noises, who first saw the apparition, and who was most frequently and most

---

[16]   Ibid., p. 97.

[17]   Ibid., pp. 129–30. Although the SPR had many female members (Eleanor Sidgwick, for example, was an influential member and its president from 1908 to 1909), at the end of the nineteenth century the council positions were almost all held by men. For example, the presidents from 1882 to 1907 included Henry Sidgwick, Balfour Stewart, Arthur Balfour, William James, William Crookes, F.W.H. Myers, Oliver Lodge, William Fletcher Barrett, Charles Richet, and Gerald Balfour.

[18]   Ibid., pp. 98–9.

[19]   Ibid., p. 105.

[20]   Francis Galton, *Generic Images* (London: William Clowes and Son, 1879), p. 3.

[21]   Campbell and Hall, p. 106.

[22]   A. Goodrich-Freer, *Essays in Psychical Research* (London: Redway, 1899), and *The Professional: and Other Psychic Stories* (ed.) (London: Hurst and Blackett, 1900).

conspicuously favoured with "phenomena"'.[23] Indeed, the presence in the house, the ghost that haunted it for the period of the investigation, *was* Miss Freer: 'the only continuity is to be found – itself not entirely continuous – of [Miss Freer]. But simply because she is a lady, and because she had her duties as hostess to attend to, she is unfit to carry out the actual work of investigating the phenomena in question'.[24] Of course, Miss Freer did carry out 'the actual work of investigation' regardless of her status as a 'lady', but it is significant here that she is marked as unfit because of her gender. This also suggests that the usual investigative methods of the SPR were masculinist, and that as a female member of the Society, Miss Freer's contributions were expected to fall into the traditional domestic category of 'hostess'. When she became a different kind of hostess, however, the hostess (or medium/host) for the apparitions at Ballechin, she adopted the role of the male psychical investigator: she appropriated power to see and document the ghosts. In assuming this male role, Miss Freer challenged the masculine authority of the SPR, but the SPR also undermined Freer's reputation by using her own ghosts against her. By discrediting her findings, the SPR suggested that her talent for seeing ghosts was all in her mind. Being too good at seeing ghosts meant that Miss Freer was to become ghostly herself in the annals of psychic research.

If we consider Miss Freer to be the ghost haunting Ballechin, then the haunted house becomes a significant metaphor for the haunted mind, particularly the female haunted mind. Critics of the female Gothic have already made connections between the haunted house and the female psyche, but they have not discussed how women's self-haunting impacts their writing and construction of the female in the ghost story.[25] Although Miss Freer has become a spectral figure in the history of the SPR, for a brief period she was a powerful female presence. Women recording psychical experiences, like Freer, as well as the women ghost-story writers I address later in this chapter, were turning to the ghost as a haunting and powerful symbol for women's disenfranchisement. Women writers, fictional and non-fictional, were aware that they were apparitional figures in constructions by the male medical community, but subversively used ghostliness as inspiration for their careers.

---

[23]   Campbell and Hall, p. 184.

[24]   Goodrich-Freer, *Alleged*, p. 84.

[25]   For discussions of connections between the haunted house and the female mind in the Gothic, see, for example, Eugenia C. Delamotte, *Perils of the Night: A Feminist Study of Nineteenth-Century Gothic* (New York: Oxford University Press, 1990). Delamotte argues that the woman's domestic space in the house both traps her and symbolizes her emotional imprisonment. Kate Ferguson Ellis, in *The Contested Castle: Gothic Novels and the Subversion of Domestic Ideology* (Urbana: University of Illinois Press, 1989), suggests that the castle represents the Gothic heroine's fear of being both confined and abandoned. Maggie Kilgour's *The Rise of the Gothic Novel* (London: Routledge, 1995) shows how Ann Radcliffe's *The Mysteries of Udolpho* (1794) articulates 'a place of confinement in which the repressed female imagination is able to escape and run riot' (p. 121). Alison Milbank argues that the female Gothic tells the story of 'an escape from an encompassing interior' (pp. 10–11) in *Daughters of the House: Modes of the Gothic in Victorian Fiction* (Basingstoke: Macmillan, 1992).

Spiritualism, Hysteria, and Writing about Ghosts

At the same time that ghosts at the fin de siècle had been adopted by the scientific, medical, and spiritualist communities as objects of study, the ghost story was also refined by the woman writer as a distinctively female form. Jessica Amanda Salmonsen argues that as much as 70 percent of ghost stories published in British and American magazines in the nineteenth century were written by women,[26] and Julia Briggs calls the end of the nineteenth century 'the high-water mark of the form'.[27] At the fin de siècle, dozens of women ghost-story writers contributed to magazines and collections of short stories, including women who have received some recent critical attention (M.E. Braddon, Amelia Edwards, Edith Nesbit, and Margaret Oliphant) and others who have been almost forgotten critically (Louisa Baldwin, Helena Petrovna Blavatsky, Emilia Dilke, Lettice Galbraith, Anna Bonus Kingsford, Mary Louisa Molesworth, and Charlotte Riddell). Even women writers not usually connected with the ghost-story genre experimented with the form, including Harriet Beecher Stowe.

The end of the century was a time of rapid social and legal change for women. In 1882 the Married Women's Property Act was passed, granting both men and women legal ownership of and rights to property, and gave women the right to litigate in their own names. In 1886 the Guardianship of Infants Act gave divorced women stronger legal rights to custody of their children. Women were becoming strong advocates for their legal identities and public selves, in particular the New Woman, who demanded that all women gain the right to education, employment, and greater social freedom. Why, then, in a period when women tried to redefine the roles they were to play in contemporary society and the coming century, was the ghost story the popular choice for the female pen? Recent work has shown that the ghost story was a popular form for the woman writer not only practically, because ghost stories promised more financial security than other genres, but also because the form was a transgressive space which allowed women to write in politically coded terms about their ghostly role in society.[28] Women were also writing ghost stories, however, because the ghost was a symbol for the ways in which they were haunted by the desire to create in a society resistant to female expression.

---

[26] Jessica Amanda Salmonsen, 'Preface', in *What Did Miss Darrington See? An Anthology of Feminist Supernatural Fiction*, ed. by Jessica Amanda Salmonsen (New York: Feminist Press, 1989), pp. ix–xiv (p. x).

[27] Julia Briggs, *Night Visitors: The Rise and Fall of the English Ghost Story* (London: Faber, 1977), p. 14.

[28] See, for example, Basham; Dickerson; Lowell T. Frye, 'The Ghost Story and the Subjection of Women: The Example of Amelia Edwards, M.E. Braddon, and E. Nesbit', *Victorians Institute Journal*, 26 (1998), 167–209; Clare Stewart, '"Weird Fascination": The Response to Victorian Women's Ghost Stories', in *Feminist Readings of Victorian Popular Texts: Divergent Femininities*, ed. by Emma Liggins and Daniel Duffy (Aldershot: Ashgate, 2001), pp. 108–25.

Despite the professional successes of the woman writer and the rise of a mass female reading public, nineteenth-century society was troubled by the conviction that writing was dangerous for women. During her mental breakdowns, Virginia Woolf was prescribed the 'rest cure' in which she was forced to remain in bed without any mental stimulation, especially writing.[29] The Gothic short story 'The Yellow Wallpaper' (1892), by American writer Charlotte Perkins Gilman, ironically shows that the boredom, depression, and insanity that the 'rest cure' was meant to prevent actually worsened these symptoms.[30] In 'The Yellow Wallpaper' the depressed narrator is forbidden to write, and it is this loss of a creative outlet which plunges her deeper into despair and madness. Writing, which women saw as both productive and curative, was often viewed as extremely dangerous by male doctors, who demanded that women stifle creativity in favour of health. Although writing was represented as a hazardous activity that brought up anxieties about controlling both the female mind and body, increased female production of the ghost story at the fin de siècle suggests that women were aware of the similarities between anxieties about the writing female and the ghosts haunting the ghost story. Perhaps in writing about ghosts, women writers were addressing their frustration with their ghostly role, not only in Victorian society, but also in serious literature. Writing 'serious' literature was hugely important to the New Woman writer's political mission, and as I will show in Chapter 6, she was anxious that writing about the supernatural might trivialize her aim to highlight women's unequal position socially, economically, and politically. At the same time, the supernatural was inspiring, allowing women writers to reach ecstatic heights of creativity.

Women's fascination with writing ghost stories in the late nineteenth century also reflects women's interest in spiritualism, where the ghost was a powerful, subversive, and inspiring emblem for female identity.[31] Indeed, women spiritualists understood the potential for ghosts to be empowering, or as Marlene Tromp has shown, how ghosts had 'transformative power':

---

[29]   S. Weir Mitchell pioneered the 'rest cure', which was influential even on Freud. See his *Wear and Tear, or Hints for the Overworked*, 8th edn (Philadelphia: Lippincott, 1897). First published in 1871, *Wear and Tear* suggested that activities like writing damaged women's physical and mental health. John Harvey Kellogg's *Ladies Guide in Health and Disease, Girlhood, Maidenhood, Wifehood, Motherhood* (London, 1890), first published in 1882, also expresses concerns about women performing 'strenuous' mental activity like reading and writing.

[30]   Charlotte Perkins Gilman, 'The Yellow Wallpaper', in *Daughters of Decadence: Women Writers at the Fin-de-Siècle*, ed. by Elaine Showalter (London: Virago, 1993), pp. 98–117.

[31]   See also Sarah A. Wilburn, *Possessed Victorians: Extra Spheres in Nineteenth-Century Mystical Writings* (Aldershot: Ashgate, 2006). Wilburn suggests that scenes of possession by spirits during a séance give women greater social power: '[i]nstead of having the ownership of private property as the necessary authorization for individual identity, a particular physical (not merely psychical) ability to become possessed is what authorizes individual expression and civic partnership according to nineteenth-century writers of mystical accounts' (p. 6).

at séances, women's access to 'the ghost, the disruption of sex and gender codes, the appropriation of "othered" identities, the use of drugs and alcohol, the blurring of identity, and the storytelling that followed them were all means of accessing "altered states", sites of intellectual, emotional, physical and spiritual refiguration.[32]

Furthermore, many women activists were also spiritualists: in identifying with ghosts, women believed they were channelling energy in order to effect social and political change. According to Alex Owen, women spiritualists gained greater social influence and autonomy while also challenging traditional understandings of class and gender. Female mediums were seen as important authorities on the spirit world, and during a séance could demonstrate sexuality and even masculinity, something that troubled Victorian expectations of female behaviour. As Owen has shown, the supposed '"innate" femininity' which women possessed 'allowed them to accede to positions of power during the séance',[33] but at the same time, the 'transgressive femininity which emerged during some of the more extraordinary materialisation séances' caused medical practitioners to see only 'the spectacle of femininity gone awry'.[34] In their understanding of spiritualism, medical men were trying to make distinctions between, as Owen terms it, 'normative womanhood' and 'pathological' womanhood.[35] The fact that spiritualists like Louisa Lowe and Georgina Weldon were incarcerated in insane asylums by their own husbands and because of their spiritualist beliefs is suggestive about how dangerous this 'pathological' womanhood was perceived to be in both the medical community and Victorian society.[36]

Significantly, the late nineteenth century was marked by an increased interest by medical men in diagnosing and researching female illness.[37] According to Elaine Showalter, the rise of both female activism and female illnesses like hysteria at the fin de siècle was indicative of male discomfort with liberated women: 'doctors ... explicitly linked the epidemic of nervous disorders [...] to *fin-de-siècle* women's ambition'.[38] Male doctors and conventional Victorian society were increasingly alarmed by the possibility that the woman would no longer be the 'angel of the house': '[f]eminism, the women's movement and what was called "the Woman Question" challenged the traditional institutions of marriage, work,

---

[32] Marlene Tromp, *Altered States: Sex, Nation, Drugs, and Self-Transformation in Victorian Spiritualism* (Albany: State University of New York Press, 2006), p. 4.

[33] Alex Owen, *The Darkened Room: Women, Power and Spiritualism in Late Victorian England* (London: Virago, 1989), p. 2.

[34] Ibid., p. 2.

[35] Ibid., p. 140.

[36] Roy Porter, Helen Nicholson, and Bridget Bennett, eds, *Women, Madness, and Spiritualism*, 2 vols (London: Routledge, 2003).

[37] Elaine Showalter, *The Female Malady: Women, Madness, and English Culture, 1830–1910* (London: Virago, 1987).

[38] Ibid., p. 121.

and the family'.[39] The figure of the woman at the end of the century had become culturally and socially unpredictable and threatening, and both her body and her mind were represented by male society as unexplored territory.

Hysteria, perhaps more than any other nervous disorder in the nineteenth century, was explicitly connected with the unknowability and uncontrollability of the female body. Sondra M. Archimedes suggests that 'potentially all of women's diseases […] could be confidently assigned to the uterus'.[40] Janet Beizer, however, points out that by the fin de siècle, the medical community was beginning to see hysteria as a 'neurological' disorder.[41] Freud's *Studies On Hysteria* (1893–1895), although preoccupied with the bodily symptoms of the illness, suggested that hysteria was a sickness of the mind. He argued that it was a result of 'psychical trauma' (which he later qualified as being almost necessarily sexual) from the patient's past.[42] For Freud,

> the splitting of the consciousness which is so striking in the well-known classical cases [of hysteria] under the form of 'double conscience' is present to a rudimentary degree in every hysteria, and that a tendency to such a dissociation, and with it the emergence of abnormal states of consciousness (which we shall bring together under the name of 'hypnoid') is the *basic phenomenon of this neurosis.*[43]

At the end of the century, medical men not only understood hysteria as an illness of both mind and body, they were also attempting to control and classify women's minds and bodies in unprecedented ways. These attempts, however, only made female identity even more apparitional and immaterial in the eyes of medical science, since neither her mind nor her body could be neatly mapped and contained.

The notion of an apparitional female identity emerging out of late Victorian medical theories on hysteria is particularly significant in terms of spiritualism. According to Roy Porter, Helen Nicholson, and Bridget Bennett 'women who underwent spiritualist experiences were readily labelled "hysterical"'.[44] As Molly McGarry has shown, '[b]oth mediums and hysterics performed and produced bodily states that at once confounded and informed men of science'.[45] In communing with ghosts, women were making themselves inaccessible to medical

---

[39]   Elaine Showalter, *Sexual Anarchy: Gender and Culture at the Fin de Siècle* (London: Bloomsbury, 1991), p. 7.

[40]   Sondra M. Archimedes, *Gendered Pathologies: The Female Body and Biomedical Discourse in the Nineteenth-Century English Novel* (London: Routledge, 2005), p. 2.

[41]   Janet Beizer, *Ventriloquized Bodies: Narratives of Hysteria in Nineteenth-Century France* (Ithaca: Cornell University Press, 1994), p. 7.

[42]   Sigmund Freud and Josef Breuer, *Studies on Hysteria*, ed. and trans. by James and Alix Strachey, *The Pelican Freud Library*, 15 vols (Harmondsworth: Penguin, 1974), III, p. 56.

[43]   Ibid., pp. 62–3.

[44]   Porter, Nicholson, and Bennett, p. 5.

[45]   Molly McGarry, *Ghosts of Futures Past: Spiritualism and the Cultural Politics of Nineteenth-Century America* (Berkeley: University of California Press, 2008), p. 126.

science, and, as McGarry suggests, both mediumship and hysteria offered them a means of 'female expression'.[46] Although medical and biological science saw women's elusive identity and connection to ghost-seeing as deeply debilitating, the next section demonstrates how women writers, like spiritualists, used the ghost as a means of accessing greater freedom of expression. Women writers saw the ghost as a symbol for their own anxieties about the role of women in the home and legally, and the ghost story as a space for political expression.

Ghostwomen, Ghostwriting

Women writers like Charlotte Riddell, Mary Louisa Molesworth, Anna Bonus Kingsford, Emilia Dilke, and Helena Petrovna Blavatsky are crucial to a discussion of women's apparitional identity in the late nineteenth century because they have themselves become ghostly in the canon. Many of these women were struggling to find a voice in late Victorian society, whether it was to advocate theosophy, like Blavatsky, or the women's trades unions, like Emilia Dilke. Furthermore, they engaged in significant ways with contemporary theories on mental science and hysteria, using ghosts and altered states like dreaming to show the ways in which women could reclaim the ghostly as a subversive symbol for female identity.

In the case of Charlotte Riddell's work, sometimes no ghosts appear in the stories and the woman herself represents the supernatural presence in the text.[47] Riddell's 'The Open Door' (1882), for example, is the story of the supposedly haunted Ladlow Hall. A young man goes to the hall to investigate a mysterious door which will not remain closed, due, village gossip says, to a ghost. The young man soon learns that the door leads to a room where a man was murdered; his will was never found. The 'ghost' is revealed to be the murdered man's living wife, who returns again and again to the property searching for the will in order to destroy it, since she knows she will inherit nothing if it is found.[48] In taking on a ghostly form the woman becomes a powerful presence in the text, not only because she protects her finances, but also because the fear she evokes gives her authority. In 'The Open Door', Riddell channels the potential supernatural fiction has for empowering women.

---

[46] Ibid.

[47] For more on Riddell writing about women writing, see, for example, Margaret Kelleher, 'Charlotte Riddell's *A Struggle for Fame*: the Field of Women's Literary Production', *Colby Literary Quarterly*, 36 (2000), 116–32, and Linda H. Petersen, 'Charlotte Riddell's *A Struggle for Fame*: Myths of Authorship, Facts of the Market', *Women's Writing*, 11 (2004), 99–115. While Petersen argues that Riddell 'reinscribes myths of female authorship [...] myths of genius and vocation, of solitude and loneliness, of domesticity and inspiration' (p. 100), she does not discuss how these myths of authorship impact on Riddell's ghost stories.

[48] Charlotte Riddell, 'The Open Door', in *Weird Stories*, new edn (London, 1885), pp. 48–103.

The women in ghost stories by Mary Louisa Molesworth demonstrate, alongside a sense of fear of ghosts, an affinity with them, a compassion and sympathy for the ghostly. Perhaps this connection is the manifestation of the subtle recognition that the ghosts they see are reflections of themselves and their own frustrated desires and ambitions. The most moving example of this alliance between the ghost and the woman can be found in Molesworth's 'Lady Farquhar's Old Lady: A True Ghost Story' (1873). Lady Farquhar and her family are staying in an old country house, Ballyreina, so that Lady Farquhar can recover her health. There, she sees the ghost of an old woman three times. She subsequently discovers that the old woman is the ghost of the eldest Miss Fitzgerald, who used to inhabit Ballyreina but whose family lost all of their fortune and were forced to move to the Continent. Miss Fitzgerald died abroad, exactly a year before Lady Farquhar inhabited the house.[49]

What is curious about the story is what it seems *not* to say about Lady Farquhar. Why is it that she feels such empathy for 'my old lady', and also so protective of her (p. 272)? When Lady Farquhar is asked by the narrator to give an account of her ghost, she speaks of it with a certain sorrowful hesitation, as if fearful of somehow betraying Miss Fitzgerald: '[a]ll that I feel is a sort of shrinking from the subject, strong enough to prevent my ever alluding to it lightly or carelessly. Of all things I should dislike to have a joke made of it' (p. 273). She also tells of her guilt and grief at not being able somehow to help the ghost:

> I cannot now describe [the ghost's] features beyond saying that the whole face was refined and pleasing, and that in the expression there was certainly nothing to alarm or repel. It was rather wistful and beseeching, the look in the eyes anxious, the lips slightly parted, as if she were on the point of speaking. I have since thought that if *I* had spoken, if I *could* have spoken – for I did make one effort to do so, but no audible words would come at my bidding – the spell that bound the poor soul, this mysterious wanderer from some shadowy borderland between life and death, might have been broken, and the message that I now believe burdened her delivered. (pp. 280–81)

The wistful, beseeching language seems to describe the feelings of both the ghost *and* Lady Farquhar. Furthermore, the anxiety about silences in this passage is suggestive about the women's silent voice politically in the late Victorian period. Finally, the notion of breaking through the 'borderland' may suggest Lady Farquhar's awakening to the fact that women are not only silent, but invisible in the nineteenth-century socio-cultural framework. Perhaps the 'borderland' represents the division between women's acceptance of their subservient role and the ushering in of the strong voices and material changes that the New Woman made towards the end of the century.

---

[49]   Mary Louisa Molesworth, 'Lady Farquhar's Old Lady: A True Ghost Story', in *The Penguin Book of Classic Fantasy*, ed. by Susan A. Williams, (London: Penguin, 1995) pp. 272–85. Subsequent references to this edition are given after quotations in the text.

Significantly, Lady Farquhar's sightings seem to coincide with a particularly unhappy or anxious time in her life, although what is making her unhappy is always kept hidden from the reader. She is at Ballyreina because 'I had not been as well as usual for some time (this was greatly owing, I believe, to my having lately endured unusual anxiety of mind)' (p. 274). She never elucidates this comment, and only refers to her nervous constitution again in order to assure the narrator that she is 'not morbid, or very apt to be run away with by [her] imagination' (p. 273). But it is during this time of anxiety that she sees the ghost, and it seems this is also the case when she is suffering from any kind of mental agitation. For example, she sees the ghost for the second and third times just after receiving a letter which is 'a very welcome and dearly-prized letter, and the reading of it made me feel very happy. I don't think I had felt so happy all the months we had been in Ireland as I was feeling that evening' (p. 279). Who sent the letter, what its contents are, and why they have made her so happy are never clear, but directly thereafter she sees the ghost. That she sees the ghost in an altered state of emotional intensity suggests that what she sees is more than a woman who, like herself, is seeking to be disburdened, but also that she is looking at herself. The ghost seems to be a projection of all Lady Farquhar's unspoken anxieties and hopes in this story (and here it is important to remember that she is also unable to speak to the ghost), manifesting itself in a reflection of inner turmoil. Lady Farquhar does seem to see the ghost, but she seems in a sense also to *be* that ghost, whose history, like her own in this tale, might reveal 'many a pitiful old story that is never told' (p. 284). Farquhar is haunted by herself here, by her own silence about her thoughts and past, as well as by the ghost of Miss Fitzgerald.

While in ghost stories like 'Lady Farquhar's Old Lady' women are haunted by themselves, writers like Anna Bonus Kingsford and Helena Petrovna Blavatsky suggest in their ghost stories that it is the act of writing which is haunting. Both women turned to ghost stories at the end of their lives, explicitly desiring these works to be representative of their lifetime achievements. Indeed, they seemed to be following a trend in ghost stories at the fin de siècle, in which ghosts appear at the moment of death. For example, 'The Story of the Rippling Train' (1888), by Mary Louisa Molesworth, is about a woman whose ghost comes to bid farewell to her lover at the very moment of her death.[50] Lettice Galbraith's 'The Ghost in the Chair' (1893) is about the appearance of a ghost at an important board meeting at the same time he is reported to be dead.[51] Louisa Baldwin's weeping woman in 'Many Waters Cannot Quench Love' (1895) appears not only at the moment of her death, but, as the reader discovers, at the same moment that her lover at sea also perishes.[52]

---

[50] Molesworth, 'The Story of the Rippling Train', in *Four Ghost Stories*, pp. 227–55.

[51] Lettice Galbraith, 'The Ghost in the Chair', in *New Ghost Stories* (London, 1893), pp. 51–66.

[52] Louisa Baldwin, 'Many Waters Cannot Quench Love', in *The Shadow on the Blind* (Ashcroft: Ash-Tree, 2001), pp. 59–67.

For Kingsford and Blavatsky, however, the ghost story became the form in which it was not the fictional ghosts who appeared at the time of death, but their own authorial personae. The supernatural became a source of comfort for these writers, which implies that not only did the ghost story allow women to express their anxieties about the plight of women more freely than other genres, but also the ghost story offered them a source of sympathetic self-identification. This self-identification was, however, modified by both an embrasure and a rejection of the affinity between ghosts and women.

Kingsford received a medical education at the Paris Faculty of Medicine, and was both a spiritualist and an influential member of the Theosophical Society. She put her spiritualism above all other aspects of her life, eventually rejecting both Anglicanism and her marriage. Most of *Dreams and Dream-Stories* (1888) was written in 1886, although it was not published until after her death.[53] According to the editor Edward Maitland, 'the publication is made in accordance with the author's last wishes'.[54] The work is a record of Kingsford's dreams, poems she claims to have written in her dreams, and stories that were inspired by or somehow relate to dreams. Kingsford explains dreaming in terms echoing those that Myers and Stead use to discuss the workings of the conscious and unconscious mind, arguing that 'the soul has a twofold life, a lower and a higher. In sleep the soul is liberated from the constraint of the body, and enters, as an emancipated being, its divine life of intelligence [...]. The night-time of the body is the day-time of the soul'.[55] The dream-life that Kingsford describes suggests that in certain somnambulic states, the separate selves within us are liberated, allowing the subservient self to have free rein. Although comparable to Stead's notion of the separations between the dominant, masculine, waking self and the suppressed, feminine, dormant self, Kingsford's dreams imply a radical re-visioning of a divided and gendered identity. Kingsford's dreams, often about powerful and mythic women leading in battle, suggest that when women access the dream or unconscious state they both confront their suppressed ambitions and desires and use these alternative states of perception as inspiration for their writing. Ghost and dream stories offer space for women to subversively adopt, and as I will suggest later, abject, the ghost as a symbol for female empowerment.

Kingsford's ghost story 'Steepside' symbolically expresses women's frustrated ambitions and explores anxieties about the act of writing. On the way to spend the holiday with friends, a young man is forced by bad weather to stay in a haunted mansion, where he sees blood rushing under his bedroom door and the ghosts of two women, one fleeing from the other across the snow. A Catholic priest later tells him that the house was once inhabited by a couple, their daughter, Julia, and Julia's

---

[53]   Anna Bonus Kingsford, *Dreams and Dream-Stories*, ed. by Edward Maitland (London: Redway, 1904), p. 9. Edward Maitland collaborated with Kingsford on many of her writings.

[54]   Ibid., p. 6.

[55]   Ibid., pp. 13–14.

maid, Virginie. Julia was supposed to marry a wealthy heir, but she secretly planned to elope with her poorer lover, Philip Brian. Meanwhile, Virginie also loved Philip and wanted to destroy his relationship with Julia. She forged Julia's handwriting in a letter to Philip, writing that the affair was over, and then informed Julia's parents about the intended elopement. Believing Julia no longer cared for him, Philip committed suicide. Philip's death sent Julia into a psychopathic rage, causing her to shoot her father in the head, strangle her mother, and finally, fling herself off a cliff. The Catholic priest learned this horrifying story from Virginie herself, who made her death-bed confession to him, and the young man was convinced that what he had witnessed in the mansion was a repetition of the dark events leading to Julia's death.[56]

Besides expressing intense female rage and thwarted desire (in the cases of both Virginie and Julia), 'Steepside' has an underlying theme of the sinister aspects of writing. When the young man first enters the haunted mansion, he reads books covered in blood. After Julia's death, Virginie goes mad and spends the rest of her life making exact copies of any written words she sees (repeating, it seems, the endless hours of practice it took her to master Julia's handwriting). The appearance of bloody books and the compulsive copying of any written text suggest that writing is somehow dangerous, and in the case of Virginie, that writing can become a punishment for past wrongs. Virginie, is after all, haunted by her own hands, penmanship, and ability to write and copy things exactly, just as much as she is by Julia's ghost. 'Steepside' expresses women writers' anxieties about the act of writing, while imagery like bloody books suggests that writing is draining for women writers, but also morbidly fascinating. Writing about ghosts is a means of symbolically articulating women's ghostly role in society and politics, but also suggests that women must turn to the ghostly and the supernatural to gain a voice. Paradoxically, women must ghost themselves to give their opinions material shape in the form of the stories.

Helena Petrovna Blavatsky, like Kingsford, also turned to the ghost story at the end of her life. A spiritualist and extremely influential theosophist with dozens of publications under her name, Blavatsky published her collection *Nightmare Tales* in 1892. The foreword to the work, written by Annie Besant, president of the Theosophical Society and political activist, expresses Besant's ideas about Blavatsky's decision to create *Nightmare Tales*:

> The world knows H.P. Blavatsky chiefly by her encyclopaedic knowledge, her occult powers, her unique courage. This little book, composed of stories thrown off by her in lighter moments, shows her as a vivid, graphic writer, gifted with a brilliant imagination. [...] The *Nightmare Tales* were written during the last few months of the author's pain-stricken life: when tired with the drudgery of THE THEOSOPHICAL GLOSSARY she, who could not be idle turned to this lighter work and found therein amusement and relaxation. Her friends, all the world over, will welcome this example of gifts used but too rarely amid the strain of weightier work.[57]

---

[56] Kingsford, 'Steepside: A Ghost Story' in *Dreams*, ed. by Edward Maitland, pp. 116–46.

[57] Annie Besant, 'Foreword', in H.P. Blavatsky, *Nightmare Tales* (London, 1892), p. 3.

Besant's perspective on Blavatsky's motivations in creating these stories was perhaps affected by the public's general unease when it came to classifying and judging women's ghost stories. The stories themselves are certainly not 'light': 'The Cave of the Echoes' and 'The Ensouled Violin' are particularly gruesome stories about jealousy, murder, vengeful ghosts, and the devil himself.[58] Although Besant dismisses the ghost story as simply an amusement genre without serious literary merit (as opposed to Blavatsky's *The Theosophical Glossary* [1892], which for Besant was a significant literary, spiritual, and intellectual endeavour), Blavatsky spent a lifetime pledging herself to theorizing about and contacting the spirit world. It seems clear that not only would she devote her 'last few months' to the supernatural, but that this would be an important, self-revelatory task.[59]

As with Kingsford, writing and its implications for female identity and ghosts is a theme in Blavatsky, in particular in her tale 'A Bewitched Life'. In this story, a young man makes his fortune in Japan in order to support his sister and her family in England. He has always been sceptical about religious faith, but when more than a year passes and he has heard nothing from his sister, he fears the worst and turns to the Japanese holy men, the Yamabooshi. He looks through a magical mirror and sees that his sister has gone mad after learning of her husband's death, and has been placed in an insane asylum; her children have been sent to an orphanage. Stricken with grief, the man refuses to ritually purify himself as the Yamabooshi suggest, and thus is doomed to foresee the deaths of everyone he meets and to relive the moments of his family's tragedy forever.[60]

The full title of the story is 'A Bewitched Life (As Narrated by a Quill Pen)', which points to the discursive framework in which the plot is working. The story opens in the bedchamber of the possibly fatally ill narrator. In a vision, she sees a man writing a tale – the hero of the story. The quill scratches on the page and the narrator discovers that she can understand its language, which she can translate, simply by listening, into the English language. Of course, this suggests that it is Blavatsky herself who is the narrator, on her own deathbed, haunted by the writings of a quill pen. Authorship here is blurred (is it Blavatsky who is writing – or Blavatsky the spiritualist medium? Is this automatic writing or the writing of the protagonist? Is the quill responsible?). But the boundaries between fiction and non-fiction are also blurred to create a cyclicality between Blavatsky's life and the text.[61] Since the protagonist can see the fate of everyone he meets, does this mean that he is foreseeing and describing Blavatsky's death?

---

[58]   Blavatsky, 'The Cave of the Echoes', pp. 68–80, and 'The Ensouled Violin', pp. 98–133, both in *Nightmare Tales*.

[59]   Besant, 'Foreword', p. 3.

[60]   Blavatsky, 'A Bewitched Life (As Narrated by a Quill Pen)', in *Nightmare Tales*, pp. 7–67.

[61]   The questions raised here are suggestive about Tatiana Kontou's theories that '[q]uestions of literary influence therefore merge with spirit possession. [...] Authorship, identity, gender and consciousness are all prone to uncanny transformations, to "spillage

Women's Ghost Stories and Abjection

At the moment that they were leaving this world for the next, both Kingsford and Blavatsky turned to writing about ghosts and about what was ghostly in writing. The affinity developing between ghostliness and women and the authors' self-identification with ghosts is equally marked by the rejection and loathing of the ghostly and the similarities between ghosts and women. For while ghost stories are a source of comfort for these writers, ghosts and the supernatural are still fearful presences in the stories. In a Kristevan sense these ghost stories are being abjected by their writers, and the horror of ghosts and ghostliness is expressed as abjection, particularly in the writings of Emilia Dilke, Charlotte Riddell, and Mary Louisa Molesworth.

The term abjection has been critically useful in discussions of the Gothic. In 'Abjection, Nationalism and the Gothic', Robert Miles suggests 'an understanding of the abject as a literary modality' in order to 'theorize horror as abjection', which allows him to 'pursue connections between horror, nationalism and the Gothic'.[62] Miles correctly points to the problem of abjection theory in terms of discussions of nationalism: 'nationalism is utterly historical' while 'Kristeva's theory [...] ahistorically deals with universal structures of the human psyche'.[63] Miles is able to situate abjection historically, however, by discussing the 'Other' as being 'coloured by nationalism' and 'how nationalism becomes part of the semiological economy of the unconscious'.[64]

While Miles situates abjection in the traditional Gothic period, I situate it at the fin de siècle, following the examples of critics like Kelly Hurley and Eric Savoy, who discuss the theory specifically in a late nineteenth-century framework.[65] The loss of self-identity, one of the greatest of Gothic horrors in the late Victorian period, is also the power of horror in Kristevan abjection.[66] The term abjection is

---

and seepage" in the texts'. From *Spiritualism and Women's Writings: From the Fin de Siècle to the Neo-Victorian* (Basingstoke: Palgrave Macmillan, 2009), p. 5

[62]   Robert Miles, 'Abjection, Nationalism and the Gothic', in *The Gothic*, ed. by Fred Botting (Woodbridge: Brewer, 2001), pp. 47–70 (p. 48).

[63]   Ibid., p. 51.

[64]   Ibid.

[65]   For example, Kelly Hurley uses *Jekyll and Hyde* to argue that both Stevenson's novel and Kristeva's *Powers of Horror* 'discuss the repulsive yet intriguing possibility of loss of self-identity' (Kelly Hurley, *The Gothic Body: Sexuality, Materialism and Degeneration at the Fin de Siècle* [Cambridge: Cambridge University Press, 1996], p. 42). See also Eric Savoy, 'Spectres of Abjection: the Queer Subject of James's "The Jolly Corner"', in *Spectral Readings: Towards a Gothic Geography*, ed. by Glennis Byron and David Punter (Basingstoke: Macmillan, 1999), pp. 161–74. Savoy discusses the abject in terms of the loss of selfhood, in which the double in James's 'The Jolly Corner' (1908) 'comes to signify, horrifically, the life without an identity' (p. 167).

[66]   See, for example, William Patrick Day, *In the Circles of Fear and Desire: A Study of Gothic Fantasy* (Chicago: University of Chicago Press, 1985), and David Punter,

thus an important theoretical tool in exploring women's ghost stories at the fin de siècle and the identity of the late Victorian writing woman.

Kristeva describes abjection as being implicitly tied to discursiveness: it is 'within the being of language,' and because of its connection to language has a specific significance to discussions about writing.[67] She describes abjection in terms that could imply both the ghostly and the ghost: 'It is ... not lack of cleanliness or health that causes abjection but what disturbs identity, system, order. What does not respect borders, positions, rules. The in-between, the ambiguous, composite'.[68] Neither living nor really dead, neither material nor immaterial, ghosts are ambiguous beings that are abject. But women at the end of the nineteenth century, themselves ghostly in their marginal positions in society, are also abject. Dickerson has already suggested the similarities between the ghost and the woman, their mutual marginal position in society, and their shared inability to be defined in a fixed or immutable way.[69] However, Kristeva's theory allows us to understand the powers of horror for the ghostliness within the female self. In discussing abjection within the subject, or self, Kristeva argues:

> [i]f it be true that the abject simultaneously beseeches and pulverizes the subject, one can understand that it is experienced at the peak of its strength when that subject, weary of fruitless attempts to identify with something on the outside, finds the impossible within; when it finds that the impossible constitutes its very *being*, that it *is* none other than abject.
>
> There is nothing like the abjection of self to show that all abjection is in fact recognition of the *want* on which any being, meaning, language, or desire is founded.[70]

Women were experiencing abjection when they wrote ghost stories at the end of the century. Women ghost-writers could not look to the public sphere (which was male-dominated) to find sympathy or understanding: they could not identify with a world that was ultimately hostile to female ambition, especially to the ambitions of the female writer. Yet when they turned inwards and began writing ghost stories as a means of satisfying the desire for recognition and identification, they found not the satiation of this desire, but rather only the desire itself. Abjection came for these women when, in turning to the ghost story, they found identification with the ghost who represented the abject.[71]

The Literature of Terror: A History of Gothic Fiction from 1765 to the Present Day, 2nd edn, 2 vols (London: Longman, 1996), II.

[67]   Julia Kristeva, *Powers of Horror: An Essay on Abjection*, trans. by Leon S. Roudiez (New York: Columbia University Press, 1982), p. 45.

[68]   Ibid., p. 4.

[69]   Dickerson, p. 11.

[70]   Kristeva, p. 5.

[71]   See also Marion Shaw, '"To tell the truth of sex": Confession and Abjection in Late Victorian Writing', in *Rewriting the Victorians: Theory, History, and the Politics of Gender*,

Writing ghost stories was an exercise in the abjection of the self, both the recognition and the abjection of a haunted self. Likewise, the female figures who appear in the stories play out the drama of abjection that occurs in the writing self. Abjection is, according to Kristeva, specifically related to the feminine, in particular to the mother. Kristeva suggests that when we are nauseated by the skin of milk and spit it out, we are spitting out the mother, abjecting her in the moment that we abject ourselves, since we *are* the mother, or at least a part of the mother that we have rejected.[72] For Kristeva, the mother is always abjected when confronted with or in place of language (which is part of a paternal, or masculine-oriented discourse) – '[t]here is language instead of the good breast. Discourse is being substituted for maternal care'.[73]

The abjection of the mother, or rather of female identity itself, causes a major conflict that occurs in ghost stories by Dilke, Riddell, and Molesworth. Paradoxically, the desire for recognition and ambition that causes these women to turn to the ghost story is also the realization that the ghosts are representations of this thwarted desire: the fictional women in the stories, written abjectly by female writers, come to represent the abjection of the mother, or rather of the female self. Ghosts alone are no longer wholly responsible for causing abjection in these stories. In addition to ghosts, women in women's ghost stories also cause abjection, in part because of their close identification with the ghostly. Furthermore, if abjection is caused by turning inwards, only to be faced with more abjection, then the writing woman is thus faced with her own abject, ghostly, and dangerous self. Kristeva has discussed the dangers that exist within the female body, in particular menstrual blood, which 'stands for the danger issuing from within the identity'.[74] Women characters themselves, and not just ghosts, are dangerous in these texts.

Emilia Dilke's 'The Shrine of Death' (1886), helps to illustrate the dangers of abjected female identity. In the story, a 15-year-old girl is desperately searching for the secrets of life. A witch tells her that only in marrying Death will she discover what she truly desires. The girl slowly pines away for the hidden knowledge that Death alone possesses, and at last the village priest advises that she should be allowed to marry him, and 'pass a night within [Death's] shrine, on the morrow it may be that her wits will have returned to her'.[75] The marriage ceremony is conducted, and the girl beholds, at last, her fearsome husband, who is bent over the pages of an open book. He compels her to look and read it, but

---

ed. by Linda M. Shires (New York: Routledge, 1992), pp. 87–100. Shaw is also interested in how writing is both abject and empowering, but overlooks the ways in which abjection is an especially important means for studying women's writing.

[72]   Kristeva, p. 3.

[73]   Ibid., p. 45.

[74]   Ibid., p. 71.

[75]   Emilia Dilke, 'The Shrine of Death', in *The Shrine of Death and Other Stories* (London, 1886), pp. 11–24 (p. 16). Subsequent references to this edition are given after quotations in the text.

she cannot decipher the writing. Frustrated, afraid, and increasingly angry, she cries: 'What shall the secrets of life profit me, if I must make my bed with Death?' (p. 23). She attempts to flee the crypt but is stopped by the 'dreadful dwellers' of the Shrine of Death (p. 23). When in the morning the villagers come to fetch her, 'she was dead; but her eyes were wide with horror' (p. 24).

'The Shrine of Death' is a significant text because it depicts awakening female sexuality and thwarted female ambition. That the girl's intense desire to know the secrets withheld from humanity begins during puberty is not coincidental. Her desire for knowledge just beyond her reach seems to mirror the stirrings of pubescent curiosity about burgeoning sexuality. The female body during this developmental period becomes unknown and unknowable, terrifying in its metamorphosis from child-body to woman-body, in which the knowledge of the self is desirable but seemingly impossible, and in which the boundaries between sexual innocence and sexual awareness are blurred. The text is littered with symbolic flowerings and deflowerings, innocence and realizations. Red roses, in particular, represent both innocence and its loss: the rose suggests the blood of the virgin and menstrual blood, the very shape and colour of the flower connoting the vagina. Roses remind the girl of her awakening desires, but also of her mother and her youth: 'she brought [the rose] close to her face, and its perfume was very strong, and she saw, as in a vision, the rose garden of her mother's house and the face of one who had wooed her there in the sun' (p. 19). When the girl walks down the stairs into the Shrine of Death, 'she heard, as it were, the light pattering of feet behind her, but turning, when she came to the foot, to look, she found that this sound was only the echoing fall from step to step of the flowers [roses] which her long robes had drawn after her' (p. 19). Here, the fallen roses sound like the footsteps of children, symbolic of the girl's lost childhood, or perhaps of the possibilities for motherhood that marriage might bring. The girl's decision to marry Death marks her ultimate deflowering, suggestive of a post-lapsarian loss of innocence but also of *le petit mort* of orgasm, the symbolic 'small death' of sexual climax that is here made literal in the consummation of marriage with Death.

What makes this story most captivating is the symbolic destruction of the girl's desire not for sexual pleasure, but for intellectual gain. The girl's desire is, ultimately, for knowledge, but when at last she is face to face with Death she finds that she cannot read the open book; regardless of her hopes and sacrifices, her wish for intellectual fulfilment is thwarted and her punishment for making such a wish is death. Many of the stories in Dilke's collection echo the themes of sexual awakening and frustrated ambition. In 'The Silver Cage', a woman grows tired of waiting for Love to come for her soul and so gives her soul to the Devil, an act which eventually causes her death.[76] In 'The Physician's Wife', a young woman marries an older physician in the hope of gaining greater scientific knowledge. She soon falls in love with the physician's young laboratory assistant, who is too cowardly to elope with her and abandons her. Wrathfully, she murders her husband and then

---

76   Dilke, 'The Silver Cage', in *Shrine of Death*, pp. 27–36.

dies alone in their castle. Her ghost haunts the castle, and by some strange curse all women who subsequently live there are destined to meet a similar fate.[77] 'The Black Veil' tells the story of an abused wife who, tired of her husband's beatings, at last gains the courage to free herself from him and murders him. However, her black mourning veil is under the control of her husband's vengeful spirit and daily grows heavier until she can scarcely move. Under the advice of a wise woman she goes to his grave to apologize, but is murdered by her husband's ghost.[78]

In *The Shrine of Death and Other Stories*, and in Dilke's subsequent collection *The Shrine of Love and Other Stories*, published in 1891, the protagonists are plagued by their hopes and ambitions.[79] The result is their downfall, and although ghosts make regular appearances in these stories, the real ghosts are the protagonists, haunted by their own dreams, desires, and actions. More specifically, when hopes and ambitions in these stories are thwarted it is almost always because they are the kind of goals that are denied women. Higher knowledge and intellectual achievement in the nineteenth century are conventionally seen as being part of the male sphere. The woman's sphere, typically a domestic one, is closed to the kinds of learning the women in Dilke's stories desire. Women for Dilke are thus haunted not simply by their own ambitions but by the very fact that they are women. The female sex in Dilke's stories is perhaps more ghostly than any of the other phantoms she conjures up.

Dilke's collections of stories are part of an emerging pattern in fin-de-siècle ghost stories written by women in which it is not the ghosts themselves which are ghostly. Instead, the female self and body is ghostly, has the power to haunt, and is expressed as abjection. The girl in 'The Shrine of Death', who is in the process of becoming a woman and whose only desire is for knowledge, is doubly abject. The want of knowledge is met in the Shrine of Death with simply want: she is unable to read the book which presumably contains the knowledge she seeks. The small village in which she grew up failed to satisfy her curiosity and lacked the imagination, sympathy, and means to bring her knowledge, but in turning to the 'impossible' she is faced only with her own unfulfilled desire, embodied metaphorically by Death. She is abjected not only by her longing for the impossible, but also by her own female identity. In seeking knowledge she is symbolically seeking the answers to her own developing 'woman's' identity. However, puberty itself as a threshold state may initiate the abject because, as Kristeva defines the term, abjection 'does not respect borders, positions, rules. [It is] [t]he in-between, the ambiguous, composite'.[80] The girl's blurred position between girl and woman makes her abject and turns the story into an allegory for female abjection: the girl's adolescence, her very self, is abjected, but in the story this translates into her literal death at the moment that she should transform into a

---

[77] Dilke, 'The Physician's Wife', in *Shrine of Death*, pp. 40–56.

[78] Dilke, 'The Black Veil', in *Shrine of Death*, pp. 79–84.

[79] Dilke, *The Shrine of Love and Other Stories* (London, 1891).

[80] Kristeva, p. 4.

woman. However, the 'motion' through which she abjects herself is thus 'within the same motion through which "I" claim to establish *myself*'.[81] The death in this story seems to signify that the abjection of female adolescence is also the birth of the female adult self. The metaphor is a cautionary one, however, since the female adult self emerges from death, suggesting that the denial or impossibility of knowledge is implicit in this emergence.

Both Charlotte Riddell's 'Old Mrs. Jones' (1882) and Mary Louisa Molesworth's 'Unexplained' (1888) use female adolescent abjection as the borderland in which ghosts can be seen and the dangers to female identity explored. The females in these stories seem either particularly wary of ('Unexplained') or prone to ('Old Mrs. Jones') nervous disorders, especially hysteria, making the link to the Victorian notion that sensitive women and female adolescents were more likely to see ghosts. Susan Schaper argues that '[w]omen's susceptibility to occult manifestations, real or imagined, was often attributed by both sceptics and believers to women's innately and distinctly feminine nature'.[82] Furthermore, she suggests that '[f]emale ghost–seeing [...] is profoundly equivocal in Victorian culture. It can serve as a testimonial to woman's highly developed sensitivity and offer her the opportunity to extend both her talent for care-giving and propensity for religion into the realm of spiritual suffering. However, ghost-seeing can also indicate psychological instability'.[83] I want to examine how ghost-seeing in these stories was 'profoundly equivocal' for the ghost-seers themselves.

In 'Old Mrs. Jones', the eponymous ghost, allegedly murdered by her husband, Dr. Jones, is said to haunt a lodging house which was once her home. Rumour says that she will not rest until her body has been found. Anne Jane, a girl who has come to work at the lodging house, is most affected by the ghost of Mrs. Jones. In fact, when she arrives, the ghost ignores all of the other lodgers and focusses solely on Anne. Eventually, in a somnambulic or hypnotized state, Anne finds the house where Dr. Jones is living under a pseudonym, and police later find Old Mrs. Jones's body. The girl's own adolescence and nerves put her at risk in this story – her female identity has, in a sense, condemned her to the seeing of the ghost and caused the abjection which forces the affinity between the ghost and the woman. However, in succumbing to the hypnotic trance state, Anne Jane is also able to grant Mrs. Jones justice. The liminal states women can reach give them authority, symbolizing their desire to attain legal and political power. The not-quite-dead and the not-quite-woman form an uneasy alliance in this story, suggesting the ghostliness that haunts female identity and the female body in this genre.[84]

---

[81]   Ibid., p. 3.

[82]   Susan Schaper, 'Victorian Ghostbusting: Gendered Authority in the Middle-Class Home', *The Victorian Newsletter*, 100 (2001), 6–13 (p. 7).

[83]   Ibid., p. 8.

[84]   Riddell, 'Old Mrs Jones', in *Weird Stories*, pp. 230–314.

Mary Louisa Molesworth's 'Unexplained' is an extraordinary, eerie piece which explores in frightening detail the dangers of the female self, especially the female adolescent self, at the moment of abjection. The story is narrated by a mother who is travelling with her two children across Germany. The family stays in Silberbach, where they are terrified by the seemingly evil inhabitants of the village as well as the village itself, but also by the fact that Nora, the woman's daughter, has witnessed a ghost.

The mother's description of Nora portrays perfectly the in-betweenness of the adolescent female:

> She scarcely looked her age at that time, but she was very conscious of having entered 'on her teens', and the struggle between this new importance and her hitherto almost boyish tastes was amusing to watch. She was strong and healthy in the extreme, intelligent though not precocious, observant but rather matter-of-fact, with no undue development of the imagination, nothing that by any kind of misapprehension or exaggeration could have been called 'morbid' about her. It was a legend in the family that the word 'nerves' existed not for Nora: she did not know the meaning of *fear*, physical or moral. I could sometimes wish she had never learnt otherwise. But we must take the bad with the good, the shadow inseparable from the light. The first perception of things not dreamt of in her simple childish philosophy came to Nora as I would not have chosen it; but so, I must believe, it had to be.[85]

The mother is attempting here to clear her daughter of any charges that nerves or hysteria would have caused her to see the ghost, but later asserts that although they were not present before the sighting, a distinct change did come over Nora afterwards:

> Nora by degrees recovered her roses and her good spirits. Still, her strange experience left its mark on her. She was never again quite the merry, thoughtless, utterly fearless child she had been. I tried, however, to take the good with the ill, remembering that thoroughgoing childhood cannot last forever, that the shock possibly helped to soften and modify a nature that might have been too daring for perfect womanliness – still more, wanting perhaps in tenderness and sympathy for the weaknesses and tremors of feebler temperaments. (pp. 188–9)

The implications of this passage are disturbing, and demand that the reader question what is really happening in this story. Why was the ghost such a shock? Was the ghost a kind of punishment or warning for Nora, so that she might conform more easily to the womanly standards her mother feared she might otherwise not have possessed? The answers to these questions seem to suggest that Nora abjects herself because she is precariously balancing between adolescence and adulthood, and this process is symbolically played out between Nora and the ghost: Nora *is*

---

[85] Molesworth, 'Unexplained', in *Four Ghost Stories*, pp. 87–226 (pp. 97–8). Subsequent references to this edition are given after quotations in the text.

the ghost she confronts in this time of intense emotional and physical change. Furthermore, the ghost is shocking because it is a warning to Nora that she must mould herself into a more acceptable womanly framework in order to better adapt to society.

Without these explanations, the ghost seems incapable of instilling any kind of fear at all. The ghost, a well-dressed young gentleman, appeared in the rooms at Silberbach, apparently because he was looking for a teacup that Nora's mother had purchased. The reader later discovers that the gentleman, who was killed by a bolt of lightning, had been buying a particular tea set for his mother that was the same pattern as Nora's mother's. The fact that the ghost was coming back from the grave to repossess a teacup makes it hard to take him seriously, but the story is genuinely unsettling, the ghost frightening because of his implications for Nora's future sense of self.

In fact, the whole story seems to captivate and chill because the reader is aware that everything that transpires within it has a life-changing effect on Nora. The story's most eerie aspect is its reconstruction of the sinister landscape of Silberbach, which is personified as a frightening, claustrophobic monster. Silberbach also seems to be peopled by menacing characters: leering, lecherous men who call attention to Nora's burgeoning womanhood and the wicked landlord who causes the mother to narrate: '[t]here seemed something sinister in his [the landlord's] words. A horrible, ridiculous feeling came over me that we were caught in a net, as it were, and doomed to stay at Silberbach for the rest of our lives' (p. 148). The threatening landscape and lascivious men are perhaps suggestive of Nora's changing emotional and physical state, respectively. Like the land, Nora's thoughts and emotions are turbulent, frightening, and dangerous. Her body, which is no longer as 'boyish' as her mother described it, is now dangerous to her because it attracts the attention of dangerous men.

'Unexplained' is a ghost story about the ghostly nature of Nora's growing up and the dangers female identity can pose to women's sense of self. Indeed, Nora becomes the ghost of herself, since she is no longer the 'merry, thoughtless, utterly fearless child' that she used to be (p. 188). Despite the fact that she seems to conform to masculine expectations for feminine behaviour, her experience is also an awakening. Indeed, in some ways her encounter with the ghost strengthens her character; for example, she now feels 'tenderness and sympathy for the weaknesses and tremors of feebler temperaments' (p. 189). The 'feebler temperaments' could refer not only to the ghosts to which she is sensitive, but also to women in the late nineteenth century who seem to have only an insubstantial political presence. Nora's newfound sympathy with ghosts and ghostly women suggests that she is alert to women's social and political invisibility, an awareness which may motivate and politicize her.

If abjection in these stories is a kind of rejection or death of the adolescent self, however, it is also, as Kristeva has suggested, an establishment or birth of the adult self.[86] Kristeva suggests that "'I" am in the process of becoming an other at the expense of my own death. During that course in which "I" become, I give birth to myself'.[87] Women writers writing about death, and about ghosts and ghostly women, were also rewriting themselves, recreating themselves in the process. The rise of the ghost story written by women at the fin de siècle marked not only the death of the woman and her reappearance as a ghost in her fiction and in social constructions about women, it also marked the ghost story as a site of the birth of the female author.

---

[86] Kristeva, p. 3.

[87] Ibid.

Chapter 5
Case Study:
Vernon Lee, Aesthetics, and the Supernatural

Introduction

> [T]he hostility between the supernatural and the artistic is well-nigh as great as the hostility between the supernatural and the logical.[1]

Vernon Lee serves as a useful case study for the examination of the fin-de-siècle fascination with and repulsion of the supernatural. Like mental scientists, spiritualists, and psychical researchers of the 1880s and 1890s, Lee impossibly wanted control of the inchoate elements of the world around her. Just as Henry James wanted to maintain 'academic neatness' in his writing, for Lee, literary genres, aestheticism, and the supernatural should all be contained within tight and impermeable boundaries. Lee seems to embody the paradox perpetuated in her writing's attempt to make material the immaterial: she herself cannot be easily accommodated within any one literary genre or period. While she is deeply focused on the past, particularly the eighteenth century, her writing is also modernist, and her literary styles, themes, and forms blur distinctions between the eighteenth, nineteenth, and twentieth centuries.

Lee's 'Faustus and Helena: Notes on the Supernatural in Art', first published in *Cornhill Magazine* (1880), was meant to act as an emphatic statement of Lee's aesthetic theories. The essay attempts to demonstrate the impossibility of portraying the supernatural in art, but, paradoxically, Lee herself constantly turned to representing the supernatural. For her, the notion of a supernatural which is based on 'abortive attempts to explain phenomena by causes with which they have no connection' is irrelevant (p. 75). The supernatural which interests Lee, 'that supernatural which really deserves the name', is 'beyond and outside the limits of the possible, the rational, the explicable – that supernatural which is due not to the logical faculties, arguing from wrong premises, but to the imagination wrought upon certain physical surroundings' (p. 76). The 'real supernatural', she argues, is 'born of the imagination and its surroundings, the vital, the fluctuating, the potent' (p. 80). For Lee, the supernatural is the impossible, irrational, and inexplicable which the imagination evokes, a supernatural that is paradoxically a product outside of nature but also within nature, for it is fashioned by the human mind.[2]

---

[1]   Vernon Lee, 'Faustus and Helena: Notes on the Supernatural in Art', in *Belcaro: Being Essays on Sundry Aesthetical Questions* (London, 1881), pp. 70–105 (p. 74). Subsequent references to this edition are given after quotations in the text.

[2]   Patricia Pulham's *Art and the Transitional Object in Vernon Lee's Supernatural Tales* (Aldershot: Ashgate, 2008) examines Lee's supernaturalism through both her art and

Lee's uneasy definition of the supernatural exemplifies the Victorian struggle to arrive at a conclusion about both supernatural events themselves and what exactly the term meant, since it could simultaneously describe natural, unnatural, and newly naturalized phenomena. Lee struggled to encapsulate the supernatural as outside of nature, at the same time locating it within the imagination – suggesting, as mental scientists did, that the mind could be unknown to itself. She also tried to show that the supernatural could not be depicted in art, but in her own writing she returned to the subject again and again.

Indeed, she was so anxious to divest art of any meaning outside of itself (especially the supernatural) that she tried to theorize a psychology of aesthetics, attempting to make the study of art a scientific project in both 'Beauty and Ugliness' (1897) and *Music and its Lovers: An Empirical Study of Emotion and Imaginative Responses in Music* (1932). This chapter explores the links between aestheticism and ghostliness that Lee repudiated and yet evoked, and demonstrates that her efforts to dissociate aesthetics and the supernatural only pushed them more closely together.

Lee's identification of both the tensions and connections between aestheticism and the supernatural situate her within a tradition in Victorian aestheticism which linked art and the Gothic. Walter Pater's seminal *The Renaissance* (1873), for example, suggests that the power of Da Vinci's 'Mona Lisa' lies in her embodiment of the supernatural, portraying her as a kind of Gothic succubus: '[s]he is older than the rocks among which she sits; like the vampire she has been dead many times, and learned the secrets of the grave'.[3] Catherine Maxwell suggests that Pater's *Mona Lisa* is here depicted as a *femme fatale*, a figure who would appear again in 'the literature of the fin de siècle', which was concerned with 'the connections between the portrait, the double, and death'.[4] Lee's work could easily fit into this kind of Gothic aestheticism; certainly, there is something of the *femme fatale*, or of Pater's *Mona Lisa*, in the painting of Medea da Carpi in Lee's ghost story 'Amour Dure' (1887), whose mouth 'looks as if it could bite or suck like a leech'.[5] However, Lee's writing complicates this kind of easy categorization and transgresses boundaries between literary genres and social identities. Women in Lee are feminine *and* vampiric (as in the case of 'Amour Dure') and they are also haunting. Female identity is unstable and evades capture: in 'Dionea' (1890), for example, Dionea escapes depiction in art, an evasion which places her alongside Lee's own anxieties about the (im)possibilities of portraying the supernatural

---

D.W. Winnicott's psychoanalytic theory of the transitional object. Pulham investigates notions of gender and sexuality in Lee to suggest that the supernatural and art offered a space for Lee's own sexual desires.

[3]   Walter Pater, *The Renaissance* (Chicago: Pandora, c. 1977), p. 125.

[4]   Catherine Maxwell, 'From Dionysus to 'Dionea': Vernon Lee's Portraits', *Word and Image*, 13 (1997), 253–69 (p. 253).

[5]   Lee, 'Amour Dure', in *Hauntings: The Supernatural Stories* (Ashcroft: Ash-Tree Press, 2002), pp. 7–30 (p. 13).

in art.[6] Lee's use of Gothic elements (such as ghosts, haunted paintings, and statues that come to life), rather than serving to exorcise the supernatural from her aesthetic writing, instead accentuate its links to the ghostly.

After all, in Lee, art kills. In 'Amour Dure' the painting of Medea da Carpi leads to Spiridon Trevka's death. In 'Oke of Okehurst; or, The Phantom Lover' (1886) Alice Oke believes she is the embodied ghost of her ancestor, based on their uncanny resemblance. Alice's husband, Mr. Oke, becomes convinced that Alice is having an affair with a ghost and, believing he sees the phantom, accidentally kills his wife.[7] Art is always the source of the ghostly in Lee's supernatural stories, and her paintings and statues are haunted despite herself. Although her stories usually offer a psychological explanation (for example, that Alice's husband's jealousy caused him to hallucinate the phantom lover), this only reinforces the contradictions in her writings: Lee's stories have ghosts, but they might just be products of the mind.

Talia Schaffer and Donald Lawler have pointed to ways in which aesthetic writers used the Gothic in order to work through their anxieties about aestheticism, their personal lives, contemporary social problems, and their own writing.[8] For Lee, however, Gothic themes and elements failed to offer a means of reconciling what she felt were the inherent differences between the supernatural and art. In a truly Gothic way, Lee's work became haunted by what haunted Lee: the fact that art could not be quantified and was both familiar and unfamiliar, uncanny in its resemblance to and difference from the supernatural.

Her stories 'Dionea', 'The Doll' (1927), and 'The Legend of Madame Krasinka' (1890) show how art and the supernatural actually intersect. Dionea, an orphan girl who washed up on the shores of an Italian village, seems to have a strange and powerful influence over everyone around her. She is depicted as a kind of lost idol or goddess whose beauty persuades the artist Waldemar to use her as his model. Although Waldemar tries to sculpt her, Dionea's beauty is intangible and cannot be contained in a work of art. Perhaps Lee here is subtly alluding to Edgar Allen Poe's 'The Facts in the Case of M. Valdemar'. Like Waldemar, Valdemar attempts to capture the uncapturable, in this case death itself, but the warning in both stories is clear: art, death, and the supernatural are elusive, transcending the human body and Lee's short story.

Although Dionea seems to fit into Lee's model of aestheticism by demonstrating that art cannot depict the supernatural, problematically, Dionea is not only a

---

[6]   Lee, 'Dionea', in *Hauntings: The Supernatural Stories*, pp. 31–49. Subsequent references to this edition are given after quotations in the text.

[7]   Lee, 'Oke of Okehurst; or, The Phantom Lover', in *Hauntings: The Supernatural Stories*, pp. 51–86. This story was first published in 1886 as 'A Phantom Lover: A Fantastic Story'.

[8]   See Talia Schaffer, *The Forgotten Female Aesthetes: Literary Culture in Late-Victorian England* (Charlottesville: University of Virginia Press, 2000), and Donald Lawler, 'The Gothic Wilde', in *Rediscovering Oscar Wilde*, ed. by C. George Sandulescu (Gerrards Cross: Smythe, 1994), pp. 249–68.

possible deity; she is art itself. The narrator comments on the fact that Waldemar treats Dionea as an object whose only purpose is to be copied: 'I could never believe that an artist could regard a woman so utterly as a mere inanimate thing, a form to copy, like a tree or flower. Truly he carries out his theory that sculpture knows only the body, and the body scarcely considered as human' (p. 45). 'Dionea' is also an inversion of the Pygmalion myth. While Aphrodite eventually transforms the statue of Galatea into a real woman and lover, in Lee's work Dionea is herself becoming increasingly more like a statue. The narrator writes: 'How strange is the power of art! Has Waldemar's statue shown me the real Dionea, or has Dionea really grown more strangely beautiful than she was before?' (p. 46). The real Dionea, at least for the male gazes of the narrator and Waldermar, *is* a work of art, a thing to be admired and copied. 'Dionea' is suggestive about the links between art and the supernatural, their slippery nature, and how both elude capture in this story.

Both 'The Legend of Madame Krasinka' and 'The Doll' express anxieties about the role of art and the supernatural in writing. In 'Madame Krasinka', the eponymous heroine sees a sketch of an old woman, known locally as Sora Lena, who committed suicide. Madame Krasinka becomes fascinated by Sora Lena's sketch and sad history, and goes to a costume ball disguised as the old woman. After the ball, she believes that she has become Sora Lena's ghost, and she unconsciously follows the woman's last movements. The story culminates in Madame Kraskinka attempting to hang herself, just as Sora Lena had done, but she is saved by the apparition of the woman herself. Madame Krasinka's endeavors to represent the supernatural in art are so successful that she becomes possessed by a ghost, and her Sora Lena disguise is so convincing that she actually becomes her.[9]

In 'The Doll' the narrator is fascinated by a life-sized doll, an exact replica of a Count's dead wife. She feels such sympathy for the doll (which seems to her deeply unhappy) that she arranges to buy it and then burns it, 'put[ting] an end to her sorrows'.[10] Once again, art in this story so completely captures the supernatural – in this case the unhappiness of a dead woman – that the art must be destroyed. Art in Lee *is* supernatural, and because of this, art in her stories always has the most power and potential. Lee's discomfort with the supernatural invading her writing is clear in the conclusion of the story: the doll must be burned and removed from the story in order to put its, and Lee's own, anxieties to rest.

Lee was haunted by her attempts, and failures, to find an aestheticism defined by materialism: she wanted an art that was materialistic, not spiritual, even when it was drawing from spiritual or supernatural themes.[11] Angela Leighton

---

[9]   Lee, 'The Legend of Madame Krasinka', in *Hauntings: The Supernatural Stories*, pp. 287–304.

[10]   Lee, 'The Doll', in *Hauntings: The Supernatural Stories*, pp. 277–83 (p. 283). This story was first published under the title 'The Image' in 1896.

[11]   In 'Resurrections of the Body: Women Writers and the Idea of the Renaissance', in *Unfolding the South: Nineteenth-Century British Women Writers and Artists in Italy*, ed. by Alison Chapman and Jane Stabler (Manchester: Manchester University Press, 2003),

argues that 'Victorian aestheticism is essentially a materialistic creed' and '[t]hat "sentiment of the body," no longer spirited into otherworldliness but fleshed with sense, pervades both English and French aestheticism. The physics rather than metaphysics of being attract the aesthete'.[12] In her discussion of ghost stories by Lee, she writes that '[i]magined as a kind of stenograph, a machine to catch voices, the ghost story provides the only means, mechanical and unbelievable, with which to manage the supernatural'.[13] Leighton further argues that in Lee's 'A Wicked Voice', Zaffirino's voice, '[t]his confusion of flesh and mechanism, the performer denatured into pure instrument, gives the song an inescapable and pervasive body'.[14] While Leighton argues that in Lee, art is haunted by its own materialism, emphasis on form, and commodity, I suggest that Lee's aestheticism, in particular her writings on music, offers an alternative vision in which art is uncanny, not because it is materialistic but because it can never achieve the materialism that Lee desires for it. Throughout her literary career, Lee was to engage in a radical and sustained embrace and rejection of the supernatural. In an examination of her aesthetic and supernatural texts, and of the links between her writing, spiritualism, and the SPR, this chapter suggests that, despite herself, Lee's was to be a supernatural aestheticism.

Lee's Ghostly Aestheticism: Haunting James, Genre, Identity, and Literary Periods

Lee was at the forefront of aesthetic writing at the end of the century, and she knew and interacted with many of the most important literary figures of her day.[15] She met Oscar Wilde in 1881 and had a mutually influential relationship with Henry

---

pp. 222–38, Angela Leighton argues that aestheticism is essentially invested in materialism. For Leighton, the relationship between aestheticism and the supernatural is problematic, suggesting that aestheticism was concerned with form and the 'intrinsic' nature of beauty, instead of what art might represent. Kathy Alexis Psomiades's *Beauty's Body: Femininity and Representation in British Aestheticism* (Stanford: Stanford University Press, 1997) centres on a similar theme, demonstrating that aestheticism had become increasingly materialistic by making the icon of aestheticism, the female body, into a form of commodification. She, too, points out, however, the connection to the Gothic, suggesting that the female body can be frightening and even dangerous, because in adopting the dress and appearance of the aesthete, a woman is also demonstrating an 'ability to take on different kinds of femininity at will, that gender itself is not an absolute category' (p. 156), a possibility which puts femininity itself at risk.

[12] Angela Leighton, 'Ghosts, Aestheticism, and Vernon Lee', *Victorian Literature and Culture*, 28 (2000), 1–14 (p. 2).

[13] Ibid., p. 4.

[14] Ibid., p. 5.

[15] Lee published prolifically on aesthetics, both in essays and books like *Belcaro* (1881), *Juvenilia* (1887), *Laurus Nobilis* (1909), *Beauty and Ugliness* (1911), *The Beautiful* (1913), and *Art and Man* (1924).

James, whom she satirized in her short story 'Lady Tal' (1892) and to whom she dedicated her aesthetic novel, *Miss Brown* (1884). Catherine Maxwell argues that in the case of Lee's ghost stories,

> James had appropriated or come to possess the stories in much more than a simple borrowing of motifs. In his own later supernatural stories he was to enlarge upon a particular feature of Lee's work [...]: the margin of ambiguity she creates in her studies so that they can be read simultaneously as both psychological studies of obsessive states of mind and supernatural occurrences. [...] in doing this, James was to complicate the distinction he makes [...] between the real and the fantastic; a complication of which Lee herself was always very much aware.[16]

That both Lee and James wrote in such a way as to confuse the real and the supernatural also suggests how genre cannot be rigidly policed: genre itself is haunted by its inability to enforce strong boundaries. Furthermore, Lee and James's relationship is suggestive about the ghostliness inherent in writing and influence. Maxwell describes their relationship as 'an extremely complicated affair of mutual acknowledgement, mutual influence, and mutual grievance'.[17] She argues that in *The Wings of the Dove* (1902), James is attacking the figure of Medea in Lee's 'Amour Dure', 'which seems to signal the way in which Lee and her writing has 'possessed' James's imagination and bears witness to the efficacious haunting of her male rival's consciousness'.[18] While this dynamic is suggestive, the fact that Lee and James were blurring the boundaries of authorship is even more significant, particularly because, as I have discussed in Chapter 1, they were so adamant that outside influence was dangerous to writing. In their ghost stories James and Lee blurred the lines between fact and fiction, haunted each other's stories, and made indistinct the separation between aestheticism and the supernatural, suggesting that in spite of their desire to neatly control and contain their writing, scenes of writing in Lee and James were always haunted.

Like her aesthetic contemporaries Pater and Wilde, Lee 'shared [the] conviction of the desirability of a perfect fusion of content and form'.[19] She believed that art could be judged only in terms of the piece of artwork itself: '[t]he goodness of the form must not be a fittingness to something outside and separate from the form, it must be intrinsic to the form itself'.[20] For Lee, we are not meant to seek in art our own emotions and desires; rather, we must appreciate what is already there. Lee's idea that art has intrinsic value extends to her discussion of art's purpose, which

---

[16]   Maxwell, 'From Dionysus', p. 267. According to Maxwell, 'Lee produced her supernatural tales long before James wrote his, a fact he acknowledged in 1890 when he thanked her for sending him a copy of *Hauntings*' (p. 267).

[17]   Ibid., p. 268.

[18]   Ibid., p. 269.

[19]   Hilary Fraser, *Beauty and Belief: Aesthetics and Religion in Victorian Literature* (Cambridge: Cambridge University Press, 1986), p. 205.

[20]   Lee, 'The Child in the Vatican', in *Belcaro*, pp. 17–48 (p. 40).

she argues is an ethical one: 'though art has no moral meaning, it has a moral value; art is happiness, and to bestow happiness is to create good'.[21] For Lee, art has benefits for the spiritual health of a person: 'by its essential nature, by the primordial power it embodies, all Beauty, and particularly Beauty in art, tends to fortify and refine the spiritual life of the individual'.[22]

In 'Notes on the Supernatural in Art' Lee emphasizes many of the aesthetic points she makes elsewhere, namely the importance of form and an understanding of art's intrinsic value in an appreciation of art. According to Lee, 'mature artists […] see only as much as within art's limits,' they are restricted by their form, and it is for this reason that they can never capture the supernatural in their art (p. 87): 'For the supernatural is essentially vague and art is necessarily essentially distinct: give shape to the vague and it ceases to exist' (p. 74). Her emphasis on form also suggests that she wants her writing to be contained in rigidly separated genres, making clear distinctions between the ghost story, for example, and historical writing. Paradoxically, she suggests that trying to give shape to the supernatural (which she does in writing ghost stories) destroys it and makes the art simply what it was to begin with: '[t]he artists were asked to paint, or model, or narrate the supernatural […] but see, the supernatural became the natural, the gods turned into man, the madonnas into mere mothers, the angels into armed striplings, the phantoms into mere creatures of flesh and blood' (p. 75). For Lee, 'art had been a worse enemy [to the supernatural] than scepticism' (p. 85), and in art '[t]he gods ceased to be gods not merely because they became like men, but because they became like anything too definite' (p. 81).

Certainly in her short story 'The Gods and Ritter Tanhuser' (1913), Lee's writing seems to effectively kill off the godliness of the gods. The story depicts the Greek gods in their retirement, and satirizes them as having embarrassing human flaws, such as arrogance, pride, and foolishness. The gods rarely live up to their reputations and even the goddess of love is undesirable.[23] This story seems to encapsulate all that Lee hoped to prove about how art and the supernatural were 'at variance' ('Notes', p. 80). The gods are not like gods because they have been confined and defined within the story, which strips them of their mysterious, immaterial power. Indeed, at the end of 'Notes on the Supernatural in Art', Lee argues that any attempt to depict the supernatural in art would always be a failure:

> Call we in our artist, or let us be our own artist; embody, let us see or hear this ghost, let it become visible or audible to others besides ourselves; paint us that vagueness, mould into shape that darkness, modulate into chords that silence – tell us the character and history of those vague beings ... set to work boldly or cunningly. What do we obtain? A picture, a piece of music, a story, but the ghost

---

[21]   Lee, 'Ruskinism: The Would-Be Study of Conscience', in *Belcaro*, pp. 198–229 (p. 229).

[22]   Lee, *Laurus Nobilis: Chapters on Art and Life* (London: Lane, 1909), p. 16.

[23]   Lee, 'The Gods and Ritter Tanhuser', in *Hauntings: The Supernatural Stories*, pp. 195–200.

is gone. In its stead we get oftenest the mere image of a human being; call it a ghost if you will, it is none. And the more complete the artistic work, the less remains of the ghost. (p. 94)

Significantly, to her list of arts which cannot capture the supernatural, Lee adds writing. Although Lee claims the supernatural can be evoked only if we keep it to ourselves 'and remain satisfied if the weird and glorious figure haunt only our imagination', she is never able to follow her own advice (p. 104). She turns again and again to the ghost story, writing over and over about the set of supernatural events she denies art can represent. Although Lee might simply be attempting to aestheticize the ghostly, using the ghost story as a form for telling beautiful and scary tales, the supernatural was actually ghosting her art, making her own aesthetics a haunted, uncanny one.

Indeed, Henry James, in a letter to his brother William, calls Lee 'uncanny', which here seems to suggest both Lee's familiarity and unfamiliarity or anxiety with her own work.[24] Ghosts in Lee are everywhere, even in her non-fiction, despite her assertion that they should stay within the confines of the imagination. 'Ravenna and her Ghosts' (1894) serves as an example. An essay on Ravenna in Italy, this non-fiction piece is curiously included in the collection of Lee's supernatural tales published by Peter Owen in 1956, *Pope Jacynth and More Supernatural Tales*. The publisher's note explains that the essay 'is not a story, although in this vignette is retold a medieval legend of the supernatural. It has been included in this volume because it is not far from the stories which make up this book'.[25] Lee's fiction and non-fiction seem to blur together, making indistinct the differences between factual and fictional accounts and demonstrating the slipperiness between genres. That ghosts populate both Lee's fiction and her non-fiction suggests that she is ghosting genre: all genres are haunted by other genres, and no genre is pure. For example, Lee's 'Ravenna' is an example of travel writing, history, ghost story, folklore, and memoir. Genre's generic differences are uncanny in their ultimate similarity, asking questions about the nature of genre itself and also straining the boundaries that Lee sets up for herself between art and the supernatural, fiction and non-fiction.

In 'Ravenna', Lee's writing demonstrates a constant evoking and exorcising of the ghosts she finds in the Italian town of that name, and her first impression of the town is immediately overshadowed by ghosts: '[a]ll round the church lay brown grass, livid pools, green rice-fields covered with clear water reflecting the red sunset streaks; and overhead, driven by storm from the sea, the white gulls, ghosts you might think, of the white-sailed galleys of Theodoric, still haunting the harbour of Classis' (p. 126).[26]

---

[24]   Quoted in Maxwell, 'From Dionysus', p. 269.

[25]   Lee, 'Ravenna and her Ghosts', in *Pope Jacynth and More Supernatural Tales* (London: Owen, 1956), pp. 124–46 (p. 125). Subsequent references to this edition are given after quotations in the text.

[26]   Theodoric is the 'King of the Goths' (p. 137), the Ostrogothic barbarian whose ghost still lingers over the Ravenna he once commanded, and who is himself haunted by the men he killed during his reign.

She then attempts to lay to rest these ghosts by emphasizing the people that live there: 'Since then, as I hinted, Ravenna has become the home of dear friends, to which I periodically return, in autumn, or winter or blazing summer, without taking thought of any of the ghosts' (p. 127). Lee continues to populate the town, attempting to dissociate it from the ghostly by suggesting that the past which might have haunted it is so remote as to be negligible:

> [t]hat is the thing about Ravenna. It is, more than any of the Tuscan towns, more than most of the Lombard ones, modern, and full of rough, dull, modern life; and the past which haunts it comes from so far off, from a world with which we have no contact. (p. 129)

But Lee inevitably resurrects the ghosts: 'Little by little, one returns to one's first impression and recognises that this thriving little provincial town with its socialism and its *bonification* is after all a nest of ghosts, and little better than the churchyard of centuries' (p. 131). At the end of her essay, she cannot resist once again haunting her text, attaching a Gothic legend, originating from Ravenna, in which every week a ghostly hunt runs through the forest in pursuit of a young woman.

Ghosts literally appear in Lee's writings, but they play more than a thematic role. As Peter Buse and Andrew Stott's *Ghosts: Deconstruction, Psychoanalysis, History* reminds us, 'modern theorists, the inheritors (and deformers) of the Enlightenment, find the trope of spectrality a useful theoretical tool'.[27] Buse and Stott cite Derrida's *Spectres of Marx* as an example of ghosts in theory, and suggest that ghosts and ghostliness are not only spectral beings but also act as abstracts for ways of conceptualizing theory, history, and identity itself. Lee's writings are haunted by her other writings, which results in a kind of ghostly meta-text about haunting itself. She is also haunted by her sense of and desire for the past, by her own self, and by her inability to exorcise from herself her anxieties about the supernatural. For while Lee desires to abject the supernatural from her work, it still influences her, coming back to haunt her in her writing. She desires to kill off the supernatural with her art as she promises in 'Notes on the Supernatural', but she cannot do it. Art, which is meant to be a matter of form, is as unfixable as the supernatural itself.

Indeed, Lee's identity also seems inchoate, as if she is unable to police her opinions about either the supernatural or her own sense of self. Critical discussions of Lee's ghost stories have focused on how Lee was conflicted about her identity by referring to the repressed or subversive sexuality which her ghosts help bring to the surface.[28] Lee's relationships with women made her identity unstable in the eyes

---

[27] Peter Buse and Andrew Stott, 'Introduction', in *Ghosts: Deconstruction, Psychoanalysis, History* (Basingstoke: Macmillan, 1998), ed. by Peter Buse and Andrew Stott, pp. 1–20 (pp. 5–6).

[28] Many critics argue that Lee's anxieties about same-sex desire emerge in her ghost stories. In '"A Wicked Voice": On Vernon Lee, Wagner, and the Effects of Music', *Victorian Studies*, 35 (1992), 385–408, Carlo Caballero argues that in 'A Wicked Voice',

of a society which would have demanded heterosexual relationships. Ambivalent sexuality haunted the mainstream society, and to a certain degree probably haunted Lee herself.[29] Particularly, Lee's supernatural stories demonstrate examples of homoerotic desire and repressed sexuality, but my interest takes this to a wider scope, examining Lee's conflicted identification with and anxieties about the supernatural, how this was reflected in her theories on aestheticism, and their relation to the fundamental contradictions in her work. I am interested in the ways in which, despite herself, her aesthetic and supernatural writings are inextricably intertwined.

Vineta Colby, whose biography on Lee 'is an attempt to read [Lee's] entire work in its fullest context – biographical, literary, and intellectual', outlines the difficulty of attempting to place into any literary category an author as diverse and as contradictory as Vernon Lee: 'In the end, Vernon Lee fits no single category. She was too late to be a Victorian, too early to be a Modernist. She was a nonmilitant

---

the singer Zaffirino represents castration, the 'erod[ing] differences [...] between the dead and the living, the past and the present, the male and the female' (p. 389). For Caballero, 'sexual mutability [...] proliferated' (p. 404) in Lee's supernatural stories. Catherine Maxwell also discusses 'A Wicked Voice', arguing that Zaffirino is decidedly effeminate, which not only indicates possible male-male desire between Zaffirino and Magnus, but also brings up questions of androgyny ('From Dionysus', p. 262). Perhaps Peter G. Christensen states this idea most explicitly in '"A Wicked Voice": Vernon Lee's Artist Parable', *Lamar Journal of the Humanities*, 15 (1989), 3–15, claiming that not only does Zaffirino represent castration, but hearing him sing has the power to turn the listener into a woman (p. 8). In *Vernon Lee: Aesthetics, History and the Victorian Female Intellectual* (Athens: Ohio University Press, 2003), Christa Zorn explores female desire in Lee, implying that Lee's 'Prince Alberic and the Snake Lady' 'favours a woman-centred and perhaps even a lesbian perspective, as suggested by [Prince Alberic's] image, which has been established as an icon of same-sex love in *fin-de-siècle* literature' (p. 156). Critics like Kathy Alexis Psomiades and Dennis Denisoff have suggested that Lee's aesthetic writings also demonstrate her anxieties about same-sex desire. In '"Still Burning from this Strangling Embrace": Vernon Lee on Desire and Aesthetics', in *Victorian Sexual Dissidence*, ed. by Richard Dellamora (Chicago: University of Chicago Press, 1991), pp. 21–41, Psomiades finds that in Lee's novel *Miss Brown*, there is evidence of lesbian desire between Miss Brown and Sacha Elaguine, a desire which Miss Brown finds both thrilling and horrifying. In 'The Forest Beyond the Frame: Picturing Women's Desires in Vernon Lee and Virginia Woolf', in *Women and British Aestheticism*, ed. by Talia Schaffer and Kathy Alexis Psomiades (Charlottesville: University of Virginia Press, 1999), pp. 251–69, Denisoff suggests, however, that Lee's aesthetic writings allow her to explore her own hidden desires: Lee [...] combined the visual genre with a feminist aestheticism [...]. Doing so allowed [her] to take essentializing artistic conventions that hindered individual exploration and self-expression and reconfigure them into literary tools of contestation for women who wished to articulate their unsanctioned emotional needs and desires (p. 251).

[29]   Terry Castle's *The Apparitional Lesbian: Female Homosexuality and Modern Culture* (New York: Columbia University Press, 1993) suggests that lesbians are socially ghostly, and that they have been neglected for so long that they seem to haunt the boundaries of mainstream society. While hers is a study of contemporary culture and her focus on an examination of the lesbian in film, she points to the idea that ambivalent sexuality can be linked to ghostliness.

feminist, a sexually repressed lesbian, an aesthete, a cautious socialist, a secular humanist. In short, she was protean.'[30] Lee herself seemed to be aware of her own unfixable nature, claiming, 'And if I contradict myself, why, I contradict myself'.[31]

Lee was obsessively fascinated with the ghosts of an idealized historic past.[32] Her *Studies of the Eighteenth Century in Italy* (1880) explores the culture and especially the music of Italy, subjects on which she fixated.[33] Although Lee's writings suggest her longing for the past, her work also places her in the centre of the literary and cultural trends of the late Victorian period. The aesthetic movement flourished at the end of the century, coinciding with the enormous rise in the popularity and production of supernatural fiction. Lee's writings blur distinctions between historical periods, literary genres, and identities, a merging of texts and themes which made her deeply anxious about the loss of control in her own writing. The next section will examine the ways in which Lee attempts to regain control: at the turn of the century and until she died, Lee focused her energies on studying, categorizing, defining, and limiting aestheticism.

'Beauty and Ugliness'

As part of her mission to police aestheticism, Lee attempted to incorporate empirical methodology into her aesthetic theories, and having read much of the writings on psychology of her day, decided that a theory of psychological aesthetics would best allow her to quantify, categorize, and make conclusions about the physical and mental response to art. In William James's *The Principles of Psychology* (1890), 'she found psychological corroboration for the connection of mind and body', a connection she had hoped to make in her psychological aesthetics.[34] Besides James, she also read Oswald Külpe's *Outlines of Psychology* (1895), and quoted him in the opening pages of the volume containing most of her works on psychological aesthetics, *Beauty and Ugliness and Other Studies*

---

[30] Vineta Colby, *Vernon Lee: A Literary Biography* (Charlottesville: University of Virginia Press, 2003), p. xii.

[31] Quoted in Colby, p. xi.

[32] See, for example, Catherine Maxwell's 'Vernon Lee and the Ghosts of Italy', in *Unfolding the South*, ed. by Alison Chapman and Jane Stabler, pp. 201–21, which 'explores the imaginative importance of the past' for Lee, whose 'writing on history, memory and association is pervaded by a form of imaginative perception and interpretation which she identifies with ghosts and ghostliness' (p. 201). In addition, Carlo Caballero suggests that Lee 'was haunted by the past. In her writing the past assumes the character of a ghost, an ineffable presence evoked by a place, a song, a picture from long ago' (p. 387). See also Peter Gunn, *Vernon Lee: Violet Paget, 1856–1935* (London: Oxford University Press, 1964). Gunn suggests that 'it was with the heightened sensibility of one in love that she approached the whole period' (p. 65).

[33] Lee, *Studies of the Eighteenth Century in Italy* (London, 1880).

[34] Colby, p. 154.

in Psychological Aesthetics (1912).[35] Külpe wrote that '[w]e may conjecture that the aesthetic feeling originates in a relation of the perceived impression to the reproduction it excites'.[36] My focus will be on the essay 'Beauty and Ugliness' co-written by Lee and her emotional and literary partner, Kit Anstruther-Thomson, a study that aims to find recordable physiological responses to the art they observed. They report that

> our facts and theories, if at all correct, would establish that the aesthetic phenomenon as a whole is the function which regulates the perception of Form, and that the perception of Form, in visual cases certainly, and with references to hearing presumably, implies an active participation of the most important organs of animal life, a constant alternation in vital processes requiring stringent regulation for the benefit of the total organism. (pp. 156–7)

During their research they posed the question, 'What is the process of perceiving Form, and what portions of our organism participate therein?' (p. 161). Using empirical research, they attempted to create 'stringent regulation[s]' for objectively studying subjective reactions to art.

The process of gathering data for the study involved examining a work of art and keeping a detailed record of any changes in the observer's body temperature, heart rate, movements, and respirations that might occur during this examination. For example, in the case of a chair that was used for the purpose of the study, Kit writes:

> in accompanying the movements connected with height [of the chair], the breathing seems limited by the limitations of the height; the breath does not rise as high as it can, but follows the rise of the eye to the top of the chair and then changes direction. There seems to be a pull sideways of the thorax, and the breath seems to stretch out in width as the balance swings across and the eyes alter their movement across the chair; then follows the expiration. (p. 165)

In the notes taken by Kit and written into theories by Lee, it seems that the body moved in accordance with the art it examined. Here, Kit and Lee were attempting to blur traditional boundaries between aesthetics and empirical science. Earlier on in the century, the borders between disciplines were more malleable, and gentlemen scholars could write scientific articles alongside educated scientists. *The Athenæum* regularly featured articles about the arts and the sciences, and there was considerable borrowing of ideas within the two fields. By the end of the century, however, the ratification of different subjects as distinct and professional

---

[35]   Oswald Külpe, *Outlines of Psychology: Based Upon the Results of Experimental Investigation*, trans. by Edward Bradford Titchener (London: Sonnenschein, 1895).

[36]   Quoted in Lee and Clementina Anstruther-Thomson, *Beauty and Ugliness and Other Studies on Psychological Aesthetics* (London: Lane, 1912), p. vi. The 1897 'Beauty and Ugliness' essay was reprinted in *Beauty and Ugliness*, pp. 153–239. Subsequent references to this edition are given after quotations in the text.

areas of focus meant that the arts and sciences were no longer amalgamated.[37] Kit and Lee's attempt to treat the two discourses as analogous was devastating to the project: their methods and evidence proved unconvincing and ultimately no scientific project could objectively assign why some art was pleasurable and some was not. Although reflective of contemporary interests in psychology to make finite the infinite (for example, the influential psychologists Karl Groos and Théodule Ribot admired their efforts), Lee and Kit's study met with an ambiguous reception. Not only did it not get as much critical attention as Lee hoped, but much of the attention it did receive was negative. Professor Theodor Lipps, a researcher in psychological aesthetics, heavily criticized their methods and results and, to make matters worse, Bernard Berenson, an art critic and writer, claimed that the women had plagiarized his own ideas.

The project was disastrous, not only because it irreparably damaged ties between Lee and Kit (who felt devastated by Berenson's accusations and exhausted by Lee's demanding work schedule), but also because it exposed the impossibility of transforming art into science. This impossibility was something Lee would never accept, and she tried to systematize art again in *Music and Its Lovers: An Empirical Study of Emotion and Imaginative Responses in Music* in 1932, a project which took her 25 years to complete. To research her topic, Lee made questionnaires which asked how people responded to music.[38] According to Peter Gunn, Lee was trying to 'explain tastes', but like 'Beauty and Ugliness', *Music and Its Lovers* could not achieve its ambition of making art appreciation an empirical science.[39]

Lee's regrets about her writings surface in the collected edition of *Beauty and Ugliness*. In the prefatory note to the essay of that name, as well as the conclusion, she apologizes and revises her earlier theories about psychological aesthetics. In the preface she writes that her own attitude towards aesthetics has changed, but that Kit's remains the same:

> I wish to point out [...] that my own present theory of Aesthetic Empathy is the offspring, or rather only the modified version, of the theory set forth in the following essay ['Beauty and Ugliness'], a theory due mainly not only to my collaborator's self-observations, but [...] to her own generalisations upon it. (p. 154)

She apologizes in the conclusion for her methodology, claiming, 'I had no standard of what constitutes psychological experimentation' (p. 352). And finally, in the conclusion, she abandons her earlier theory altogether:

---

[37] For more on the emergence of disciplinarity in the late nineteenth century, see Amanda Anderson and Joseph Valente, eds, *Disciplinarity at the Fin de Siècle* (Princeton: Princeton University Press, 2002).

[38] See Lee, *Music and Its Lovers: An Empirical Study of Emotion and Imaginative Responses to Music* (London: G. Allen & Unwin, 1932).

[39] Gunn, p. 229.

In short, the plural pronoun employed by me in *Beauty and Ugliness* meant
not *we two collaborators*, but *we, all mankind*, or at all events all mankind
capable of formal aesthetic preference. […] I really thought that everybody was
'we.' It was only when, reading Lipps […] that I gave up the belief that the
phenomena described by my collaborator must necessarily be taking place in
some subconscious region of my own self. (pp. 352–3)

Her rejection of her own ideas and her betrayal of Kit suggest the unhappiness
and the sense of failure with which Lee seemed to regard much of her aesthetic
writing. Her comments in *Juvenilia* (1887) act as a foreshadowing to the later
problems she would face in her aesthetic writing. In the book, Lee dismisses
aesthetical questions as youthful, happy ones which have no real profound
meaning. She writes: '[w]e were happier first. Decidedly, that is what I have been
insisting all along. But while we were happy other folk were wretched and this
convenient division of property and class cannot be kept up for good.'[40] This
statement seems to be a gentle reminder to herself that she too, must move on, and
that the empirical aesthetics she is seeking cannot be found.

Critics like Kathy Alexis Psomiades and Diana Maltz have analyzed 'Beauty
and Ugliness' in terms of the homoerotic tension between Kit and Lee and what
that says about Lee's aestheticism.[41] In contrast, Colby dismisses the significance
of 'Beauty and Ugliness', stating that Lee 'squandered valuable creative energy
in measuring physiological reactions to specific works of art'.[42] Lee's desire and
inability to quantify art, however, marks a fundamental crisis in Lee's aesthetics,
placing her alongside the psychological thought of her day and, she would have
been horrified to recognize, the psychical research of her day as well.

---

[40]   Lee, *Juvenilia: Being a Second Series of Essays on Sundry Aesthetical Questions*
(London: T.F. Unwin, 1887), p. 9.

[41]   When, for example, Kit and Lee observe a statue, Psomiades writes that '[a]esthetic
experience is thus based in lesbian desire, both Vernon's desire for the embodied Kit, and
both women's desire for the statue's revelation' ('Still Burning', p. 31). Aesthetic experience
in Lee is a physical one, argues Psomiades, in which bodies can gaze on other bodies
intellectually, artistically, and sexually. In 'Engaging "Delicate Brains": From Working
Class Enculturation to Upper-Class Lesbian Liberation in Vernon Lee and Kit Anstruther-
Thomson's Psychological Aesthetics', in Schaffer and Psomiades, eds, *Women and British
Aestheticism* (pp. 211–29), Diana Maltz also argues for a kind of sexually voyeuristic reading
of the 'Beauty and Ugliness' experiments: '[t]he museum gallery was in fact a social arena
where Anstruther-Thomson used her body to titillate an audience of female, upper-class
devotees. […] In Anstruther-Thomson's hands, this program […] became instead a lively,
liberatory forum for an aristocratic lesbian elite' (p. 213). See also Phyllis F. Manocchi,
'Vernon Lee and Kit Anstruther-Thomson: A Study of Love and Collaboration between
Romantic Friends', *Women's Studies*, 12 (1986), 129–48. Other critics have also made
connections between aestheticism, homoeroticism, and homosexuality. See, for example,
Richard Dellamora, *Masculine Desire: The Sexual Politics of Victorian Aestheticism*
(Chapel Hill: University of North Carolina Press, 1990).

[42]   Colby, p. 335.

Lee and the Ghosts of the SPR

Like other late Victorian thinkers, Vernon Lee wanted to make quantifiable the unquantifiable. She was looking for evidence to explain art, just as the SPR was looking for evidence to explain supernatural phenomena. Lee was openly dismissive of the ghosts of the SPR, and in the preface to *Hauntings* (1890) she writes:

> Hence, my four little tales are of no genuine ghosts in the scientific sense; they tell of no hauntings such as could be contributed by the Society for Psychical Research, of no spectres that can be caught in definite places and made to dictate judicial evidence. My ghosts are what you call spurious ghosts (according to me the only genuine ones), of whom I can affirm only one thing, that they haunted certain brains, and have haunted, among others, my own and my friends.[43]

Lee was interested, not in the materialistic nature of ghosts, but rather in the ghosts that sprang from the imagination. Her language, however, betrays her real interest: while spurious is defined as 'illegitimate' and 'counterfeit', the term also means 'of material things'. Lee wants her ghosts to be false, fictional, and imaginary, but she also wants them to be tangible. She defines ghost in the following terms in 'Notes on the Supernatural in Art':

> By *ghost* we do not mean the vulgar apparition which is seen or heard in told or written tales; we mean the ghost which slowly rises up in our mind, the haunter not of corridors and staircases, but of our fancies. [A ghost is] a vague feeling we can scarcely describe, a something pleasing and terrible which invades our whole consciousness, and which, confusedly embodied, we half dread to see behind us, we know not in what shape, if we look round. (pp. 93–4)

Again, her language suggests she is conflicted about what she means by ghosts. While she argues that a real ghost 'rises up in our mind', she also suggests that ghosts are 'confusedly embodied'. Indeed, this phrase aptly expresses Lee's perplexity, not only about the nature of her ghosts, but also about the nature of the supernatural in her writing: she confusedly wants ghosts to be embodied while insisting that they escape materialization.

Furthermore, Lee sharply dissociates herself from the SPR by suggesting that their only interest was in the physical evidence that could be obtained about ghosts instead of in the ethereal, imaginative ghosts that she assures us preoccupy her in her own ghost stories. The SPR was indeed invested in putting mediums to rigid scientific tests in order to discover whether they were fraudulent or not: mediums were placed in cabinets with their hands bound, and sometimes sitters held onto lengths of string attached to the mediums' hands, which would allow them to detect any movement. Searches of the séance room were conducted to rule out the use of fraudulent devices. Spiritualists also conducted these tests to prove to sceptics that the manifestations were real. The search for material evidence proved

---

[43]   Lee, *Hauntings*, 2nd edn (London: Lane, 1906), p. xi.

problematic and paradoxical for spiritualists, however, since spiritualism arose in part as a reaction against the materialism of Victorian science. Spiritualism was meant to gift a new kind of spirituality to a society that had largely lost all faith, and that felt depressed by the materialism of modern society.

Despite their supposed distaste for materialism, however, spiritualists and psychical investigators alike were connecting the spirit back to the physical body of the medium. Spiritualists demonstrated this in their delight in full-form materializations at séances, where ghosts, when touched or pinched, always felt very fleshy. The investigators were also interested in the fleshliness of séances, since often in the darkened séance room touch was the only perceptible phenomenon. As SPR member Frank Podmore comments, 'It is then, upon this unexercised and uneducated sense of touch that the investigator at a dark séance has to rely almost exclusively, not merely to inform himself of what feats are being performed, but also to guard against the medium's complicity in the performance'.[44]

The SPR was interested not only in finding evidence to support a genuine spiritual experience, but also in bodily responses to haunting events. Just as Lee and Kit had attempted to monitor the ways in which the body responded to a work of art, the SPR was interested in the reactions of the body in a ghostly encounter. SPR member Charles Richet, for example, describes how the body reacts when it is in the presence of a ghost: '[t]he arrival of phantoms is nearly always heralded by a vague sensation of horror, the feeling of a presence coinciding with a cold breath'.[45] Richet also offers a bodily explanation for spiritual phenomena such as table-turning. He argues that while sometimes the movements are inexplicable,

> in most of these cases, though not in all, these movements are to be explained by the unconscious movements of the subject. His muscles can be seen to contract, and as the least pressure will cause a table in unstable equilibrium to move, no other cause can reasonably assigned either for table-movements or automatic writing.[46]

Careful measurements of the body show that it is simply muscular action that controls the table or the pen, and for Richet, the body holds the key for unravelling the mysteries of spiritualism.

Spiritualist William Crookes's experiments suggested that the body could prove whether a medium was genuine or fraudulent. President of the SPR in the 1890s, Crookes is best known for his experiments with mediums, and in particular for his researches on the mediumship of Florence Cook and her spirit guide, 'Katie King'. He published his séance notes from the 1870s in the SPR Proceedings for 1889. In one séance with Florence Cook, he attached to her body 'an electrical

---

[44] Frank Podmore, *The Rise of Victorian Spiritualism*, ed. by R.A. Gilbert, *Modern Spiritualism: A History and Criticism*, 8 vols (London: Routledge, 2000), vol. VII, pp. 197–8.

[45] Charles Richet, *Thirty Years of Psychical Research: Being a Treatise on Metaphysics*, trans. by Stanley de Brath (London: Collins, 1923), p. 567.

[46] Ibid., p. 401.

apparatus, called a galvanometer, by means of which the young woman became part of a mild electric circuitry'.[47] When Florence was attached to this contraption, Crookes assumed she could not impersonate a materialized spirit without producing tell-tale fluctuations in the galvanometer readings.

While his experiments were often criticized for their inconsistency and poorly thought-through methodology, and while, as in the case of Florence Cook, 'Katie King' materialized without his machine registering any change, Crookes nevertheless demonstrates the interest psychical researchers took in monitoring the body when it was confronted with spiritual phenomena.

While Lee and Kit were attempting to blur the boundaries between art and science in 'Beauty and Ugliness', psychical researchers were attempting to make the spirit and the material overlap. But here a problem arose: a close identification between the spiritual and the body was a materialistic creed. Janet Oppenheim argues that nineteenth-century philosophy was uneasy about the relationship that should exist between the mind and the body:

> The problem was [...] significant to spiritualists and psychical researchers, for the independent existence of mind was, of course, an essential part of their argument against materialism. Whether dubbed mind, soul, spirit, or ego [...] such an entity distinct from brain tissue was requisite to rescue man from a state of virtual automatism, a mere bundle of physical and chemical properties. [...] Thus spiritualists and psychical researchers alike found themselves drawn to the infant study of the human mind from a scientific perspective.[48]

Like the psychical researchers and the spiritualists, Lee was caught between the body and the mind, between her monitoring of bodily responses to art and the elusive nature of art itself. She was also caught between the logic she believed existed in art and the illogic and impossibility of the supernatural that she claimed art could not capture, and yet from which she constantly drew inspiration in her own writing. Unlike the psychical researchers and spiritualists, however, Lee was striving for the materialism they were seeking to escape. But although she would have liked to exorcise her demons, Lee was only successful in being haunted by the ghosts who would not leave her writing alone.

Despite the fact that she contended her ghosts were not those of the SPR, the fact that she was haunted by ghostliness in her writing suggests that her interest, just like the SPR's, was also in the ghosts of the mind. Indeed, as Oppenheim suggests, by the end of the century the SPR, other psychical researchers, and spiritualists were all increasingly interested in what was ghostly within the field of mental science. The ghosts of the SPR were sometimes the ghosts that 'haunted certain brains', suggesting that, to an extent, Lee's ghost stories were working

---

[47]   Janet Oppenheim, *The Other World: Spiritualism and Psychical Research in England, 1850–1914* (Cambridge: Cambridge University Press, 1985), pp. 345–6.

[48]   Ibid., p. 207.

within the same framework as the mental scientists of her day.[49] That her ghost stories have been connected by recent critics to Henry James's psychological thrillers strengthens the idea that her ghosts of the mind were not simply those of an imaginative past, but were also influenced by contemporary psychology.

Lee explores this notion in 'The Hidden Door' (1886). In this humorous Gothic tale, the satirically named Decimus Little (a man 'accustomed to think of himself as connected with extraordinary matters, and in some way destined for an extraordinary end') believes he has discovered the Secret Chamber of Hotspur Hall.[50] Little accidentally opens the chamber and, believing he has released a terrible curse, flees the house in despair, only to learn later from a housekeeper that the 'chamber' is actually where the servants dry the linen.

Here the only explanation is psychological: Little's overwrought imagination has been his undoing. Significantly, Little is a man fascinated by psychology: he 'read[s] about delusions in Carpenter's *Mental Physiology*' (p. 327), and is 'considerably interested in [...] the Society for Psychical Research' (p. 322). 'Carpenter' refers to William Benjamin Carpenter, one of spiritualism's harshest opponents. He believed that all spiritualist phenomena could be dismissed as delusion on the part of the witnesses, the result of unconscious action, or by some other natural explanation. His straightforward psychological text, *Principles of Mental Physiology* (1874), was strongly critical of any belief in the supernatural.[51]

'The Hidden Door' is the only one of Lee's supernatural stories for which she offers no supernatural explanation. Curiously, though, it is the man of empirical science, who studies contemporary psychology and is 'open to arguments and evidence on all points', who believes he has come face to face with the supernatural (p. 322). Despite Carpenter's warnings that supernatural events are the result of delusion, Little fully deludes himself, for here there really are no ghosts. Perhaps in denying the supernatural to the man interested in the SPR, Lee is reiterating her distaste for the sorts of ghosts that the society was seeking, but it is also suggestive that she and Little share a similar fear of and desire for the ghostly: both try to abolish it, but find it haunting even in the most unlikely places.

Lee characterizes Little as someone who might be inclined to see a ghost: he has a strong imagination, a belief in his own extraordinary fate, and an uneven temperament (he is given to 'impulses of lawlessness' [p. 322]). However, she also denies him the one element that she lends to all of her other stories: art. While in most of her supernatural stories, ghosts emerge from paintings ('Amour Dure', 'Oke of Okehurst') or music ('Winthrop's Adventure', 'A Wicked Voice'), no ghosts emerge here at all, and this lack of the supernatural is due to the lack of art. Lee's aesthetics are built, not on the appreciation of the intrinsic nature of

---

[49]   Lee, *Hauntings*, p. xi.

[50]   Lee, 'The Hidden Door', in *Hauntings: The Supernatural Stories*, pp. 321–8 (p. 321). Subsequent references to this edition are given after quotations in the text.

[51]   William Benjamin Carpenter, *Principles of Mental Physiology* (London: Routledge, 1993).

art's form, but instead on her desire to reject the supernatural while simultaneously writing about it. 'The Hidden Door' is both a ghost story and a humorous short story, a story which reaches through genres and explodes Lee's idea that art has the power to contain only what is intrinsic to itself. Lee spent a lifetime attempting to capture and put limits on beauty, never realizing that the supernatural could redefine beauty by freeing it from the impossible, and often arbitrary, restrictions Lee imposed on herself in works like 'Beauty and Ugliness'.

Lee's Phantom Wicked Voices

Lee loved music more than any other art form, and wrote about it from her 1880 publication of *Studies of the Eighteenth Century in Italy* onwards. Quoting Pater, she summarizes her view of music in 'Impersonality and Evolution in Music' (1882):

> 'All arts,' Mr. Pater has suggestively said, though perhaps without following to the full his own suggestion – 'All arts tend to the condition of music;' which saying sums up perfectly my own persuasion that the artistic element of all arts, which in each is perplexed and thwarted by non-artistic elements, exists in most unmixed condition in music, because music is in reality much less connected with life and its wants and influences than any other art.[52]

Lee's aesthetics demanded that music, like any other art form, should have no meaning outside itself, and that its value was due to its intrinsic essence: 'music is [...] the most formal and ideal of all arts, unique in the fact that the form it creates resembles and signifies nothing beyond itself'.[53]

Despite this assertion, music in Lee is always associated with that which lies outside of its form: the supernatural and the uncanny. In 'Impersonality and Evolution in Music' Lee reviews psychical researcher Edmund Gurney's *The Power of Sound* (1880), which is concerned with the psychical aspects of sound and music. Gurney's book suggests that music, and in particular the voice, has an intense and powerful influence over its listeners that is comparable to telepathic or mesmeric forces, a link Victorians would have readily made, particularly at the end of century, with the popularity of *Trilby*. Gurney suggests that

> given fairly adequate conditions, the immediate power of one being over the feelings of another seems at its maximum in a case where no external tools or appliances are involved, where nature and art appear one, where phenomena of absolute beauty can be presented as though part of the normal communication of man to man, and where in addition the use of the familiar words heightens the naturalness of address, and completes the directness and spontaneity of the effect. Many will attend when addressed in this way whose lives would

---

[52] Lee, 'Impersonality and Evolution in Music', *Contemporary Review*, 42 (1882), 840–58 (p. 857).

[53] Ibid., p. 856.

otherwise lie wholly apart from the influence of beautiful and pure emotion. In the midst of this normal sad remoteness the effect of song on the masses is like a glimpse of infinite spiritual possibilities; and owing to the fewness of the moments where even the suggestion of a universal kinship in lofty sentiment appears possible, such occasions seem to have a very singular and impressive significance in human life.[54]

Gurney argues that music transcends conventional perceptions, suggesting that listening to music is a supernatural experience. Lee may have felt uneasy with this aspect of Gurney's theory, since later in *Music and Its Lovers* she was trying to limit perception to music, not give it 'infinite spiritual possibilit[y]'.

Lee did agree in many ways with Gurney's belief in the power of the voice, however, and she asserted that music depended on the performer.[55] The fact that she herself evokes the uncanny in music shows that she was both fascinated by and anxious about the ways in which the supernatural could destroy rational control over art:

I believe […] there exist musical forms common to all the composers of a given epoch, forms which they slightly alter and rearrange without removing the sense that such forms have been heard before, even as when we see a hitherto unknown member of a family whose face gives us the sense of having been seen before.[56]

Composers can unwittingly compose music they have never listened to before, but which they have always intrinsically heard, an occurrence which plagues Winthrop in 'A Culture Ghost: or, Winthrop's Adventure' (1881). Winthrop cannot decide whether the music that haunts him is the work of his own hand or that of the ghost of a composer he witnessed in Italy, and exclaims, 'Of course, I either composed it myself or heard it, but which of the two was it?'[57] Winthrop's experience with music is an uncanny one in which the music is both deeply intimate with him, and also a horrible unknown.

In 'A Wicked Voice' (1890), a version of 'Winthrop's Adventure' which Lee rewrote in *Hauntings*, Magnus is also eternally haunted by a music which may or may not be his own: '[m]y head is filled with music which is certainly by me, since I have never heard it before, but which still is not my own, which I despise and abhor: little trippings, flourishes and languishing phrases, and long-drawn, echoing cadences'.[58] For Magnus, both eventualities are disastrous: if the music is composed

---

[54]   Edmund Gurney, *The Power of Sound* (London: Smith, Elder, 1880), p. 475.

[55]   Lee, 'Cherubino: A Psychological Art Fancy', in *Belcaro*, pp. 129–55 (p. 151).

[56]   Lee, 'Impersonality', p. 851.

[57]   Lee, 'A Culture Ghost: or, Winthrop's Adventure', in *Hauntings: The Supernatural Stories*, pp. 251–76 (p. 255).

[58]   Lee, 'A Wicked Voice', in *Hauntings: The Supernatural Stories*, pp. 87–105 (p. 104). This story was first published in French as 'Voix Maudite' in 1887. Subsequent references to this edition are given after quotations in the text.

by a ghost, then he may be haunted forever, but if he has composed a piece that he despises, then it suggests there is also something uncanny about his very self. Part of Magnus is so unfamiliar to him that it writes horrible, haunting music.

In fact, Lee's own experience with singers and music is uncanny, for in *Studies on Eighteenth Century Music* she writes about pieces that she has never heard and perhaps never will hear, but that she still feels familiar with: 'it was a feeling of mingled love and wonder at the miracle of the human voice, which seemed the more miraculous that I had never heard great singers save in fancy'.[59] She is fascinated by the uncanny possibility of writing about music that is both absent and present, material and immaterial.

In her essay on music aesthetics, 'Chapelmaster Kreisler: A Study of Musical Romanticists', Lee also touches on themes of the uncanny in music. Her discussion revolves around Johannes Kreisler, a character in E.T.A. Hoffmann's novels.[60] It is significant that Lee chooses Hoffmann's work as a means of articulating her ideas about music, particularly because Freud later uses Hoffmann in his definition of the uncanny (pp. 219–56). Although Lee's essay predates Freud's theory of the uncanny, she seems to be aware of what Freud would term the uncanny in her understanding of music and, like him, seems to perceive the uncanny elements in Hoffmann's stories. Despite herself, Lee is evoking the strange and dark sides of music. Both 'Winthrop's Adventure' and 'A Wicked Voice' present a haunting voice which has the power to kill and which defies capture as it both eludes and pursues Winthrop and Magnus.

In 'A Wicked Voice' the composer, Magnus, is haunted, not by the materialization of a ghostly voice, but rather by its failure to materialize. In fact, very little in this story seems to fully materialize. The singer, Zaffirino, has a voice which could belong to a man or a woman, and he is physically reminiscent of both an innocent boy and a *femme fatale* (p. 97):

> That effeminate fat face of his is almost beautiful [...] I have seen faces like this, if not in real life, at least in my boyish romantic dreams, when I read Swinburne and Baudelaire, the faces of wicked, vindictive women. Oh Yes! he is decidedly beautiful, this Zaffirino, and his voice must have had the same sort of beauty and the same expression of wickedness ... . (p. 91)

Zaffirino's sexual ambiguity is reminiscent of descriptions of Trilby, whose enormous voice complicates her femininity, but his own blend of feminine and masculine characteristics suggests that he is not quite materializing as any gender

---

[59] Lee, 'Introduction: For Maurice: Five Unlikely Stories', in *Hauntings: Supernatural Stories*, pp. 177–91 (p. 187).

[60] Lee, 'Chapelmaster Kreisler: A Study of Musical Romanticists', in *Belcaro*, pp. 106–28. Johannes Kreisler is a character in some of Hoffmann's novels, for example, *Kreisleriana* (1813). Kreisler is a tormented musical genius.

at all.[61] Indeed, his very image is elusive, for when Magnus sings, the portrait of Zaffirino 'keeps appearing and disappearing as the print wavers about in the draught' (p. 92). Despite the fact that the portrait seems visually ghostly, it represents the only real material vestige of Zaffirino. The fact that Magnus destroys the portrait, however, ripping it and throwing it into the canals of Venice, suggests that like Lee, Magnus is haunted by his inability to produce materialism in his art.

After all, every time Magnus tries to compose and to create new art, he is haunted by a melody which is just out of reach, rendering him incapable of writing anything at all:

> as soon as I tried to lay hold of my theme, there arose in my mind the distant echo of that voice, of that long note swelled slowly by insensible degrees, that long note whose tone was so strong and so subtle.

> There are in the life of an artist moments when, still unable to seize his own inspiration, or even clearly to discern it, he becomes aware of the approach of that long-invoked idea. A mingled joy and terror warn him that before another day, another hour have passed, the inspiration shall have crossed the threshold of his soul and flooded it with rapture. (p. 94)

Ironically, Magnus's long-awaited inspiration is not even his own composition, but instead Zaffirino's, whose voice he can never quite catch anyway. He hears 'a ripple of music' (p. 94), but then, almost before he has registered it, 'the sounds had ceased' (p. 95). Magnus increasingly becomes obsessed with this voice, which is always just out of reach and which he can never quite hear: 'My work was interrupted ever and anon by the attempt to catch its [the voice's] imaginary echo' (p. 96). Significantly, Magnus himself uses the word 'imaginary', as if he knows he is searching for something that will never fully materialize.

Indeed, even in the fleeting moments when he does hear the voice, his body seems to dematerialize and becomes as ghostly as the voice that haunts him. As he hears the singing he says, 'A faintness overcame me, and I felt myself dissolve' (p. 95). When he hears the voice again he notes that he is becoming as immaterial and ethereal as the voice itself: 'it seemed to me that I too was turning fluid and vaporous, in order to mingle with these sounds as the moonbeams mingle with the dew' (p. 103). Magnus becomes more and more spectral, and by the conclusion he is only a ghost of the composer he was before, unable to think of any music save the 'ghost-voice which was haunting me' (p. 96). He confesses that 'I am wasted by a strange and deadly disease. I can never lay hold of my inspiration' (p. 104), and begs for Zaffirino to sing once more (p. 105).

---

[61]   Critics like Caballero and Christensen have discussed the homoerotic implications of Zaffirino's sexual ambiguity. See *Art and the Transitional Object*, in which Pulham also discusses the ambiguity of gender in Lee's 'A Wicked Voice', although she focuses on the ways in which the castrato is symbolic of the maternal voice haunting Lee's fiction.

But Zaffirino, it seems, will not materialize anytime soon, since he was never fully materialized anyway. Magnus sees him twice, but the first time is in a dream (p. 93), and the second time is from a distant gallery which resembles 'a dark box in a half-lighted theatre' (p. 102). From this room, suggestive of the darkened rooms of the spiritualist séances, Magnus sees a ghost indeed, but the ghost is so spectral that by the time he throws open the door upon the spirit, Zaffirino has disappeared into the ether and the room 'was as bright as a mid-day, but the brightness was cold, blue, vaporous, supernatural' (p. 104). At the end of the story we are left only with the supernatural, and Magnus is left with only the haunting traces of a voice he will never be able to exorcise.

In 'Impersonality and Evolution in Music', Lee explores the anxieties she later expresses in 'A Wicked Voice' about how music can evade categorization as being either a material or immaterial art form. Although she credits music with perfection because of its adherence to form, she contradicts herself by explaining how music *cannot* be contained by form: 'music is an art of emotional material, of material intimately connected with the realities of things, and of abstract, ideal form – of form unlike anything outside itself of which we have any experience'.[62] Art is both realistic and abstract, material and able to escape materialism, a matter of form but also a form which in its description is like nothing we have experienced before – in effect, the musical form she seems to be talking about (despite herself) is supernatural. Music is also supernatural in 'A Wicked Voice', and in this case the voice is compared to the devil himself: 'For what is the voice but the Beast calling, awakening that other Beast sleeping in the depths of mankind, the Beast which all great art has ever sought to chain up, as the archangel chains up, in old pictures, the demon with his woman's face?' (p. 88). The 'demon with his woman's face' is suggestive about Trilby, her sexual ambiguity and her mesmeric voice, but Magnus also seems to be describing Lee's own desire and inability to 'chain up' the supernatural in art, just as he himself is never able to harness the voice to compose the perfect piece.

That music and the voice in Lee are directly connected to the supernatural, and in particular to the immaterial, suggests that the voice in Lee was uncannily disembodied. Zaffirino is never seen to sing, and his voice comes only from empty gondolas or from the night air, never from a physical body. Spiritualism was also very much concerned with what was ghostly in the voice. Although these ghosts were not the literary ones of Lee's writing, their example is instructive in examining the conflicts between the spiritual and the material that conflicted Lee. The voice, or 'direct voice' as it was called in the séances of the spiritualists, was a disembodied one much like Zaffirino's, a spectral ghost speaking from nowhere. Alex Owen describes the spiritual experiences of the Theobald family, who heard 'the voices of spirit children not uttered through the medium's mouth but manifesting themselves

---

[62]   Lee, 'Impersonality', p. 843.

independently throughout the room'.[63] William Crookes describes manifestations of various sounds during the séance, which to him can only be the result of intelligent disembodied spirits within the room: 'the sounds to which I have just alluded will be repeated a definite number of times, they will come loud or faint, and in different places at request; and by a pre-arranged code of signals, questions are answered and messages are given with more or less accuracy'.[64]

Lee's ghosts are never really fleshed out, never fully embodied, but instead only a glimpse, a voice, a painting, or a sculpture. Lee's aestheticism wrestles with its own materialistic claims, only to come to the uneasy conclusion that art is as haunted with her failed empiricism as her ghost stories are. Both the supernatural and art in Lee are slippery, for she simultaneously attempts and fails to give them a body. Lee's supernaturalism is forever just out of reach of materialism, and her wicked voices never fully materialize, making them perpetually uncanny.

Still, Lee would have wanted her ghost stories to be Leighton's 'machine to catch voices'.[65] After all, when she and John Singer Sargent saw an eerie portrait of the castrato singer Farinelli in 1872, she wrote, 'what would we not have given if some supernatural mechanism had allowed us to catch the faintest vibrations of [Farinelli's] voice'.[66] But for Lee there was always something more desirable about the unattainable, a desire that extends to her flawed attempts to quantify art, and 'a longing for the unattainable, with the passion only unattainable objects can inspire'.[67] Indeed, she was always more fascinated with what was just out of reach, asking in 'Notes on the Supernatural', '[w]hy do those stories affect us most in which the ghost is heard but not seen? ... Why, as soon as a figure is seen, is the charm half-lost?' (p. 94).

In her discussion of ghosts and visuality in Thomas Hardy's poetry, Catherine Maxwell suggests that 'all portraiture has a link with death, [and] the silhouette has even a stronger relation in that it figures absence more graphically, so that, where the subject of representation is in fact dead, the silhouette becomes a shade of a shade'.[68] A shade, also a synonym for a ghost, offers 'the sensitive observer the opportunity of projecting more freely his or her own memories, impressions, fantasies and associations into the charged blank space of the silhouette; that is to say, it sums up what is important to the observer'.[69] Maxwell's analysis of

---

[63] Alex Owen, *The Darkened Room: Women, Power and Spiritualism in Late Victorian England* (London: Virago, 1989), p. 89.

[64] William Crookes, 'Notes of an Enquiry into the Phenomena Called Spiritual', in *Researches in the Phenomena of Spiritualism* (London: J. Burns, 1874), pp. 81–102 (p. 87).

[65] Leighton, 'Ghosts', p. 4.

[66] Lee, 'For Maurice', p. 183.

[67] Ibid., p. 187.

[68] Catherine Maxwell, 'Vision and Visuality', in *A Companion to Victorian Poetry*, ed. by Richard Cronin, Alison Chapman, and Antony H. Harrison (Oxford: Blackwell, 2002), pp. 510–25 (p. 515).

[69] Ibid., p. 515.

absences, which are more intimate, more telling than presences, is true of Lee's ideas about writing and the supernatural and her desire for what is not there, as well as of her tendency to project the ghostly into everything she wrote. Of music, Lee wrote, '[t]here is nothing stranger in the world than music; it exists only as sound, is born of silence and dies away into silence, issuing from nothing and relapsing into nothing; it is our creation, yet it is foreign to ourselves'.[70] Lee's writing worked in much the same way: the supernatural was always the uncanny double of her aestheticism.

---

[70]   Lee, 'Chapelmaster', p. 106.

Chapter 6
Balancing on Supernatural Wires: The Figure of the New Woman Writer in Sarah Grand's *The Beth Book* and George Paston's *A Writer of Books*

Introduction

> 'I was talking to a fellow only to-day, who is in a publisher's office, and he was telling me the sort of thing that the public wants. He says they don't care about all that – what d'ye call it? – analysis, and if you want to make money you should write a historical romance with lots of fighting in it, or something in the supernatural line with Biblical characters like Miss – I forget what she calls herself.'

> 'My dear Tom,' cried Cosima, aghast. 'I couldn't do such things to save my life, and I wouldn't if I could.'[1]

In this scene from *A Writer of Books* by George Paston (Emily Morse Symonds), Tom suggests that writer Cosima attempt more popular lucrative fiction; she is horrified, not just by the idea of compromising her work to suit the mass market but also, significantly, by the idea of writing *supernatural* fiction. Although we are alert to Cosima's discomfort with 'selling out' and her wish that 'she should never need to write for money, but only for name and fame', it is less clear why the thought that writing about the supernatural in particular threatens her (p. 210). When the literary market of the 1890s was flooded with novels about the supernatural, such as *The Picture of Dorian Gray*, *The Beetle*, and *Dracula*, why did New Woman writers of the same period appear to turn away from this popular trend and invest themselves in realistic fiction depicting women's anxieties about marriage and their struggles for financial independence? Were the differences between Gothic supernaturalism and fiction about the plight of women at the end of the century as distinct as they seem?

Like other women writers characterized in fin-de-siècle New Woman fiction, Cosima desires to write realistic novels in order to portray 'all sides of life and all sorts of conditions of men' (p. 45).[2] New Woman writers themselves, for example

---

[1]   George Paston, *A Writer of Books* (Chicago: Academy Chicago Publishers, 1999), p. 209. Subsequent references to this edition are given after quotations in the text.

[2]   See, for example, Sarah Grand's *The Beth Book* (London: Virago, 1980), in which Beth vows that she will write about 'the normal – the everyday' (p. 373), believing that

Sarah Grand, also firmly believed in the importance of writing accurate depictions of real life: '[t]o be true to life should be the first aim of an author, and if one deals with social questions one must study them in the people who hold them'.[3] Indeed, the genre's political agenda and its use of realism have been the key focuses of most critical writing on New Woman fiction. The notion of being 'true to life' is problematic, however, since it is not clear whether Grand means true to the conventional expectations for women (such as being a good wife and mother) or true to the New Woman cause of seeking alternatives to marriage, domesticity, and financial dependence. Grand may also imply that women should be 'true to life' in another capacity: they should be keenly alert to the paranormal insights and flickerings of the supernatural that haunt their supposedly realistic fiction.

Cosima's horror of the supernatural, which both fascinates and repulses her, and the robust endorsement of realism in her writing seem representative of the seriousness of purpose with which New Woman writers viewed their own vocation. If they turned to popular fiction (sensational, supernatural, decadent) and explored its possibilities, they were concerned that they could jeopardize their claim to legitimate social commentary. However, New Woman writers were radically ambivalent about the supernatural, both disturbed by and irresistibly drawn to the subject. Despite their firm defence of realism and scorn for popular fiction, New Woman writers *did* test the potential of supernatural themes like telepathy, inspiration by unknown agencies, and scenes of Gothic horror. While Ann Heilmann has noted that New Woman writers employ the supernatural in their writing, she has interpreted the supernatural, and in particular the Gothic, as peripheral to their central concern of political realism.[4] Far from being secondary to the New Woman's interest in evoking women's oppression, however, the supernatural actually subverts generic expectations and is an alternative political tool to realism.

This chapter contends that in New Woman fiction the act of writing is itself a supernatural act; in other words, in scenes of writing, women display extrasensory perception, which enhances their literary talent but is also physically and emotionally debilitating. While Chapter 4 suggested that women identified with the ghosts of their ghost stories and were haunted by their own female identity,

---

in this way she can learn more about human nature and help her readers to find greater happiness (p. 374). Subsequent references to this edition are given after quotations in the text. Mary, in Ella Hepworth Dixon's *The Story of A Modern Woman* (Peterborough: Broadview, 2004), also believes in the importance of writing about real life: 'I can't help seeing things as they are, and the truth is so supremely attractive' (p. 147). Subsequent references to this edition are given after quotations in the text.

[3]   Sarah A. Tooley, 'The Woman's Question: An Interview with Madame Sarah Grand', in *Sex, Social Purity and Sarah Grand: Journalistic Writings and Contemporary Reception*, ed. by Ann Heilmann and Stephanie Forward, 4 vols (London: Routledge, 2000), I, pp. 220–29 (p. 220). First published in the *Humanitarian* (1896).

[4]   Ann Heilmann, *New Woman Fiction: Women Writing First-Wave Feminism* (Basingstoke: Macmillan, 2000).

this chapter suggests that women writers are haunted by the destructive nature of the writing process and the driving ambition to write. Yet when women write in New Woman fiction, they also gain access to mesmeric and ecstatic trance states, which ultimately leads to exceptional writing. As Roger Luckhurst has suggested, 'trance, nervous sensitivity, and supernormal affinity were as much a part of the self-conception of the New Woman as the conservative attack on "neurasthenic" feminism', which suggests that the New Woman found connections to trances and supernatural awareness both intellectually and socially empowering and debilitating.[5] While in earlier chapters I discussed the ways in which Sherlock Holmes and women ghost-story writers found access to trance states empowering, in New Woman fiction trance states are dangerous and intoxicating.[6] I have chosen Sarah Grand's *The Beth Book* and George Paston's *A Writer of Books* for my discussion because they are representative texts of the figure of the woman writer in New Woman fiction. These texts need urgent re-examination in order to illustrate how supernatural concerns impact upon a discussion of feminism in fiction in the 1890s.

A consideration of *The Beth Book* and *A Writer of Books* reveals that scenes of writing are figured in the supernatural terms of telepathy or automatic writing, in which authorship and agency are contested sites of power. The uncanny nature of this writing, in which the words seem to be both familiar and alien simultaneously, is perilous, since during the act of composition female protagonists court the unknown that exists beyond masculine definitions of selfhood. Writing forces them to go to the very edge of their imaginative powers and to the brink of ill health, where they hover until they rein themselves in at the last to complete their novels. It becomes apparent that the ritual of writing in these novels is an initiatory activity. Though hampered by intense loneliness, poor health, and crippling self-doubt, they use the writing process as a means of ambiguous self-discovery and development.

Whereas my previous chapters have discussed negotiations between machines and typists and the mesmerist and mesmerized, in New Woman fiction anxieties about sites of power are internalized into the figure of the woman writer, who is doubled and divided between her private artistic ambition and her negotiations with the literary marketplace and between being a dependent wife and an independent author. In an examination of Grant Allen's *The Type-Writer Girl*, this chapter returns to discussing the connection between women, automaticity, and new

---

[5]   Roger Luckhurst, *The Invention of Telepathy: 1870–1901* (Oxford: Oxford University Press, 2002), p. 226.

[6]   In 'Marie Corelli's Magnetic Revitalizing Power', in *Victorian Literary Mesmerism*, ed. by Martin Willis and Catherine Wynne (Amsterdam: Rodopi, 2006), pp. 183–202, Alisha Siebers examines how trance states can lead to 'rich, authoritative writing' in Corelli (p. 184). At the same time, Siebers suggests that in *A Romance of Two Worlds*, Corelli questions whether trance states offer 'transcendence' or evoke 'gothic thematics such as imprisonment, relinquishment of will, and loss of self' (p. 184). In New Woman fiction, I argue, trance states are both ecstatic and entrancing, and it is this combination of the dangerous and the delightful that leads to good writing.

technology to suggest that in New Woman fiction, technology allows women to adopt either passivity or authoritative agency in the writing process. Furthermore, women writers depicted in New Woman fiction are shown to be uncannily able to succeed in a social environment hostile to female ambition: women writers transgressively adopt and turn to their advantage the very systems and discourses which often undermine or frustrate them. For example, Cosima in *A Writer of Books* both engages in and speaks out against censorship in her writing, disguising her political ideas as conventional literary tropes. I suggest that her achievement, becoming a popular writer and a writer with a serious political voice, is itself an uncanny position for a woman writer at the fin de siècle. Ultimately, in New Woman fiction, writing becomes the uncanny means by which female authors can attain heightened consciousness and fresh opportunities both in their art and in life.

The Ecstasy of Writing: Grand's *The Beth Book* and Extrasensory Perception in New Woman Fiction

Sarah Grand's Beth in *The Beth Book* is possessed with the desire for expression, but for her writing is always an ambiguous means of articulating her views. In part, this is because Beth is deeply anxious about the origin of influence in her writing. Indeed, she describes her writing as a kind of telepathy, in which the source of her creativity is unknown: '[t]hings come into my mind, but I don't think them and I can't say them' (p. 178). Beth suggests that while she knows what she wants to articulate, the language she needs to express herself is elusive and the act of writing cannot adequately express her thoughts. Her mind also seems undemarcated, almost as if there is something illicit or taboo about her ideas, which can enter and leave her channels of mind unbidden. The notion that she 'can't say them' reinforces the possibility that these are forbidden thoughts, but it is also suggestive about the voice of the New Woman in society, who was 'saying' things that were controversial to hear about, such as the changing role of women and male sexual vice.[7] The fact that she does not think them either is threatening

---

[7]   The New Woman was giving voice to the controversial issue of male sexuality and vice, subjects which were 'unspeakable' in polite society. Texts like Emma Frances Brooke's *A Superfluous Woman* (1894), Ménie Muriel Dowie's *Gallia* (1895), and Sarah Grand's *The Beth Book* and *The Heavenly Twins* (1893) were particularly concerned with the male role in the spread of syphilis. Many New Women were furious that the Contagious Diseases Acts of 1864 and 1866 meant that prostitutes could be detained in hospitals and were blamed for the spread of disease, while men were free to visit prostitutes and possibly infect their wives (and by extension, their children) with syphilis. While New Woman fiction began after most of these acts were repealed, they used the Contagious Diseases Acts as a platform to express their anger at the wrongs against women in marriage. For more on the New Woman's reaction against male sexual vice, see, for example, Emma Liggins, 'Writing Against the "Husband-Fiend": Syphilis and Male Sexual Vice in the New Woman Novel', *Women's Writing*, 7 (2000), 175–95.

to Beth because their irrational appearance in her mind conflicts with the rational thinking she, as a figure in a New Woman text, ought to uphold and demonstrate.

In a discussion of the poetry she has written, Beth makes distinctions between this kind of telepathic verse that appears in her mind and the material poetry that she would compose were she to make writing a serious enterprise:

> At least, I didn't make it up, it just came to me. When I make it up it'll most likely be quite different. It is like the stuff for a dress you know, when you buy it. You get it made up, and it's the same stuff, and it's quite different, too, in a way. You've got it put into shape, and it's good for something. (p. 178)

Again, because Beth is anxious about writing which she channels from an unknown origin and which impresses itself upon her mind, she implies that when she 'really' writes, her writing will be as concrete, domestic, rational, useful, and material as the dress she uses in her analogy. Beth's decision to be more dismissive of the irrational writing that 'comes to her', and to concentrate on the realistic writing that she 'makes up', echoes the desires of the New Woman writer both to write realistically and to use writing to voice the internalization of cultural taboos at the end of the century. At the same time, however, writing is uncanny for Beth, the product of her own words and the creation of another agency, and regardless of her desire to edit out the unknown words that come to her, she cannot write without the influence of that initial, unknown agency.

Beth is both repelled by her telepathic perception, and its potential to damage her claims to serious realist and political literature, and imaginatively drawn to it. While she embraces realism, eventually publishing a work of non-fiction and vowing 'I should avoid the abnormal' in her writing, it is the abnormal which fascinates her (p. 373). Significantly, Beth does not write that she will avoid the abnormal, but rather that she 'should', suggesting that she is aware of the realistic literary conventions she feels pressured to follow, yet conflicted about whether or not she will apply them to her own writing. Furthermore, the notion that she 'should' avoid taboo ideas suggests that it is male editorial policies which put her under pressure to uphold a writing style which adheres to popular conventions. Despite this, the abnormal becomes an irresistible temptation for her, and she delights in taking the risk of writing about the illicit and in plunging herself into writing which is enticing and scandalous because of its 'abnormality'.

Beth is tempted again by the 'abnormal' when, in a moment of wakefulness, she reaches for a novel, 'a shilling shocker [...]. The story was of an extremely sensational kind, and she found herself being wrought up by it to a high pitch of nervous excitement' (p. 436). The novel is exhilaratingly terrifying for her, and when, while reading, she hears the cries of a dog that her husband has been experimenting upon, she imaginatively transforms the scene into one of Gothic horror.[8] She kills the dog to spare it, but 'she felt the stronger for a brave

---

[8] See also Heilmann, *New Woman Fiction*, which discusses this passage in depth in Gothic terms (p. 93). Many New Women were opposed to vivisection. Significantly, in

determination, and more herself than she had done for many months' (p. 437). Her actions suggest that not only does re-imagining the scene of the vivisectionist table as a moment of Gothic suspense give her the courage to save the animal from further suffering, but also passing through the terrors of witnessing the dog's pain enables her to rediscover herself: she regains some of the confidence and moral strength she feels she has lost since her marriage to Dr. Maclure. Indeed, she is symbolically vivisecting herself, dissecting and criticizing the aspects of her life which threaten to undermine her ambitions. While the process of putting herself 'under the knife' is painful, it is also a necessary operation which will motivate and inspire her literary endeavours.[9]

Beth is most often confronted with moments of extrasensory perception, however, when she is writing, and it is these scenes of writing that ultimately act as initiatory processes. For Beth, writing is both a torment and a triumph, causing her poor mental and physical health but also giving her a purpose and usefulness that she lacks in her stifling, unhappy marriage. She insists to herself, almost desperately, 'I shall succeed! I shall succeed!' suggesting that for her, writing is an ambiguous means of achieving personal success (p. 398). While she is aware of the potential benefits of her success, she also discovers that the cost of writing is high:

> Now the things she did not care about she began to do with a rush, so as to get to her writing. She wanted to be always at that; and the consequence was a wearing sensation, as of one who is driven to death, and has never time enough for any single thing. (p. 423)

The fact that she is 'driven to death' is suggestive of the intoxicating nature of her writing process: she is both energized to write and worn away with the effort to continue writing.

The phrase 'wearing sensation' is also significant, making ambiguous the negative effects of her writing. Although the phrase implies she is overworked, the term 'wearing' suggests that she is performing identity and taking on a guise in order to write. The term 'sensation' is defined as '[a] condition of excited feeling produced in a community by some occurrence; a strong impression (e.g., of horror, admiration, surprise, etc.) produced in an audience or body of spectators and manifested by their demeanour', and 'an event or a person that "creates a sensation"'. Her writing may exhaust her, but it also creates a sensation,

---

'Gynaecology, Pornography, and the Antivivisection Debate', *Victorian Studies*, 28 (1985), 413–37, Coral Lansbury argues that part of the problem within discussions of vivisection was that the medical community treated women like the animals they were vivisecting (she refers particularly to gynaecology and theories of the female body).

[9]   See also Oscar Wilde, *The Picture of Dorian Gray* (Harmondsworth: Penguin, 1970) in which Wilde also uses vivisection as a symbol for self-investigation: '[Lord Henry] had begun by vivisecting himself, as he had ended by vivisecting others. Human life – that appeared to him the one thing worth investigating' (p. 66).

suggesting that the public hungers after the kind of writing that is emotionally and physically draining to produce. Furthermore, the use of the word 'sensation' suggests that Beth herself is inspired by this destructive process of writing: it fills her with 'horror' and 'admiration' for her accomplishments. Finally, the term 'sensation' also evokes the notion of Victorian sensation fiction. Although most popular in the 1860s, sensation fiction had a revival at the end of the century, and dealt with many of the themes Beth is both anxious to avoid and drawn to represent in her writing (and also the kind of fiction which thrills her late at night), such as the Gothic and moments of psychological suspense.[10]

Beth herself is aware of the contradictions in the ways in which she perceives her writing, and suggests yet another contradiction, that the process of writing is comparable to being in love: 'She had the same warm glow in her chest, the same sort of yearning, half anxious, half pleasant, wholly desirable' (p. 181). The comparison between being in love and writing is radically ambiguous, however, since Beth has never really been in love. Perhaps Beth's 'yearning' is for literary influence or patronage, rather than for emotional attachment. Furthermore, comparing her heightened perception to something as sentimental and as conventional as love suggests again the ongoing conflict Beth faces between her delight in reaching the very limits of herself and her desire for the rational and the tame. Her writing process is 'half anxious, half pleasant, wholly desirable', but it is not love that brings her these conflicting sensations, but rather the ecstasy of inspiration. Ecstasy is defined as 'the state of being "beside oneself", thrown into a frenzy or a stupor with anxiety, astonishment, fear or passion' and 'the state of trance supposed to be a concomitant of prophetic inspiration; hence, Poetic frenzy or rapture'. Beth is filled with both 'anxiety' and 'passion' when she writes, sending her into a deep trance state which is simultaneously exalted and fearful.[11] Timothy Clark suggests that '[i]nspiration is held to blur conceptions of agency: the writer is possessed or dispossessed and may undergo an extreme state of elation'.[12] The suspension of agency during her writing is an anxious time for Beth, and yet both her elation and her possessed/dispossessed body are essential to her writing process: it is the ecstasy, the fear, *and* the rapture which propel her to write.

---

[10]   Wilkie Collins's *The Moonstone* (1868; New York: Knopf, 1992), like Grand's *The Beth Book*, is also concerned with heightened perception, although Collins focuses on the trance state in mesmerism, rather than other kinds of extrasensory perception like telepathy.

[11]   Late nineteenth-century writers were interested in the ways in which inspiration and the trance state could be closely linked, and how inspiration was both a destructive and a rapturous moment. In Joris-Karl Huysmans's *Against Nature* (1884), trans. by Robert Baldick (London: Penguin, 1959), Des Esseintes is inspired by art which evokes feelings of both horror and delight. For him, inspiration in art is 'the feverish desire for the unknown, the unsatisfied longing for the ideal, the craving to escape from the horrible realities of life, to cross the frontiers of thought, to grope after uncertainty' (pp. 114–15).

[12]   Timothy Clark, *The Theory of Inspiration: Composition as a Crisis of Subjectivity in Romantic and Post-Romantic Writing* (Manchester: Manchester University Press, 1997), p. 3.

Beth uses her writing to help her overcome her depression and feeling of helplessness and isolation in her marriage: 'during the writing of [her book] she enjoyed an interval of unalloyed happiness, the most perfect that she had ever known. [...] The terrible sense of loneliness, from which she had always suffered [...], was suspended' (pp. 423–4). Indeed, it is when she discovers a room of her own, secret from her husband, where she can write without interruption, that she is able to begin to make decisions about her future career. In addition, this initial freedom from her husband is what eventually gives her the courage to leave him entirely. Although the 'unalloyed happiness' she feels about her writing is only momentary (the phrase itself suggests that her happiness is too perfect, or too good to be true), and although writing fills her equally with despair and elation (she describes how it 'had ceased to be a pleasure, and become an effort to express herself in that way' [p. 517]), this suggests once again that it is the intoxicating highs and lows of the activity that fascinate her, just as it is the entrancing nature of the telepathic which inspire her. But this destructive nature of writing is ultimately conducive to Beth's creative process, for it allows her to exist at the very limits of herself and gives her the multiplied perception which inspires her writing.

For example, when Beth is at the seaside she spontaneously composes a poem which is the product of some unknown agency or telepathic power that Beth cannot define. The poem is later shown to be a foreshadowing of her Aunt Victoria's death, suggesting that Beth harnesses psychic power, but it also demonstrates how closely Beth's creative life and the supernatural are intertwined. Once again, Beth's writing is here figured both as telepathic and as automatic writing, both processes in which agency is suspended. Indeed, in a description of the moments in which things 'come to her', Beth says, 'I have to hold myself in a certain attitude – not my body, you know, *myself* – hold myself in suspense as it were, or suspend something in myself, stop something, push something aside' (p. 213). The 'suspense' she speaks of seems to allude not only to a moment of horror and to psychological states, as if she is in suspended animation or a somnambulic or mesmeric trance, but also to the trance state of ecstasy. She seems to write in a liminal state of horror and rapture, the trance allowing her to balance between the two in order to find inspiration. In this moment of fearful exhilaration she is putting aside literary conventions and social and masculine expectations, pushing away her rationale in order to most freely express herself. While her words suggest that she is pushing aside her own consciousness, allowing for another self or consciousness to enter and use her body as a medium for its message, there is also the suggestion that that other consciousness is Beth's most revelatory one.

Certainly, Beth's method of writing is suggestive of the automatic writers of the spiritualist séances, who believed they were transcribing the messages of the dead, displacing their own conscious minds to allow for the entry of the words of the spirits. Beth's writing process is also reminiscent of the theories posited by researchers of the SPR, who, like F.W.H. Myers, suggested that phenomena such as automatic writing could be the result of a 'secondary self' or unconscious thought. The fact that the discourse on automatic writing is linked to mental science and pre-Freudian theories of the subconscious, and the idea that Beth's telepathy may be a result of her 'secondary self', implies that her unconscious thoughts are spilling onto the page in order that she reclaim herself from the role of repressed and oppressed woman in the late nineteenth century. This process of writing gives her access to her innermost thoughts, which suggests her writing is the expression of the anger she feels at her unequal situation in marriage and society.

While Beth is consciously aware of the wrongs of her husband, and of society in condoning such a marriage, this theory explains, in part, why Beth finds writing to be such an emotionally intensive experience. After writing until she is both mentally and physically exhausted, '[w]riting became a rage with [Beth]', suggesting that writing is a cathartic exercise for her, a means for her to articulate her frustration with a society which is hostile to women's ambition (p. 424). The term 'rage' is also, however, suggestive of inspiration, since it is defined as '[p]oetic or prophetic enthusiasm or inspiration'. Furthermore, the expression 'all the rage' (according to the OED in use since 1785) suggests that writing is also a trend: for Beth, writing is an outcry against thwarted ambition, a driving inspiration, and a fashionable accessory. The notion of her writing as a trend suggests that she wants her writing to become 'all the rage': she craves the potential to write a bestseller.

As I have suggested, Beth's writing gives her greater confidence and ultimately allows her to separate from her husband. Although this writing comes at a cost to health and happiness, Beth is one of many women writer figures in New Woman fiction possessed with extrasensory perception, which grants the gift of great literary talent. Cosima in *A Writer of Books*, for example, uses her power to help her edit her work:

> by the aid of the sixth sense which she may be said to have developed, she perceived passages crying out for improvement, skeleton characters who needed flesh upon their bones, and scenes into which by a touch she could infuse the warmth, the colour, the perfume of life. (p. 232)

While Cosima seems to be able to magic paper to life, Hester in *Red Pottage* (1899) serves as another example of a woman writer in New Woman fiction who uses her paranormal insights to enhance her writing. Hester sees her heightened perception as a means to bring her outside of herself and closer to a godly power: 'Hester spoke brokenly with awe and reverence of her book, as of some mighty presence, some constraining power outside herself'.[13] While these women writers stand on the edge of a power greater than themselves, they continue to desire to express themselves, not merely for literary greatness, which they all in varying degrees achieve, but also to create outside of normal human possibility and to be consumed by inspiration.

---

[13] Mary Cholmondeley, *Red Pottage* (London: Virago, 1985), p. 335. Subsequent references to this edition are given after quotations in the text.

To an extent, women writers in New Woman fiction are able to harness moments of heightened perception in order to bring them to the brink of the inaccessible, the beautiful, and the other-worldly. George Paston, for example, suggests that Cosima has a masterful control over her writing:

> Her imagination had now become a well-tuned instrument upon which she played with a virtuoso's hand. Like Gautier, she could launch her sentences into the air, knowing that they would fall upon their feet like cats, her brain was almost painlessly delivered of *le mot juste*, and her characters had blood in their veins and marrow in their bones. (p. 206)

Here, Cosima has not only fine-tuned her writing skills, but in doing so she also gains access to another level of her consciousness. The comparison of the musical virtuoso and the writer is apt in a discussion of multiplied perception, particularly when, as Phyllis Weliver points out, '[m]usic was an important component of mental science and a central metaphor for explaining and conceptualizing theories of consciousness'.[14] Not only was music used 'in Victorian writings about mesmerism, hypnotism, the development of double consciousness, double personality, memory and theories of identity',[15] but the violin, particularly, was connected to mental science: 'the metaphor of vibrating strings was an extremely popular means by which to explain how mesmerism worked'.[16] Cosima's mind is like the vibrating strings of the violin, enabling her to reach other powers within herself, whether they comprise the secondary or unconscious self or even mesmeric power.[17]

Despite their access to powers of extrasensory perception, writing nearly destroys the physical and mental health of writers like Beth and Cosima. Indeed, Cosima states that she writes *almost* painlessly, suggesting that the process is painful indeed. In an earlier passage Cosima writes, 'Literature is the most lonely of all trades [...] The writer works in silence and privacy, spinning his plot, making his effects, wrestling with his difficulties, with none to see, none to help, none to applaud' (p. 95). The word 'trade' here also suggests that a writer must face deprivation and give up the social, and domestic aspects of her life as a trade-off for a heightened literary experience.

---

[14]   Phyllis Weliver, *Women Musicians in Victorian Fiction, 1860–1900: Representations of Music, Science, and Gender in the Leisured Home* (Aldershot: Ashgate, 2000), p. 8.

[15]   Ibid.

[16]   Ibid., p. 64.

[17]   Thomas Hardy's 'The Fiddler of the Reels', in *The Short Stories of Thomas Hardy* (London: Macmillan, 1928), pp. 401–19, is also concerned with the links between mesmerism and the violin. In the story, Car'line both fears and desires the mesmeric sounds of Mop's violin: '[a]fter the first moments of paralyzed reverie the familiar tune in the familiar rendering made [Car'line] laugh and shed tears simultaneously' (p. 414).

Writing is often as much a torment for women writers as it is a pleasure, for while at times Beth finds no occupation which gives her greater happiness, at other times writing for her is obsessive and compulsive:

> [s]he would spend hours over one sentence, turning it and twisting it, and never be satisfied; and when she was at last obliged to stop ..., she went with her brain congested, and her complexion, which was naturally pale and transparent, all flushed and blotched with streaks of crimson. (p. 425)

Her 'congested' brain suggests that she is not only 'full', but overloaded: she has gorged on her writing and it has sustained her, at the same time consumingly, overwhelmingly filling her. Like a vampire drains blood, Beth drains herself of energy, and also like a vampire, she is overcharged because of it.[18]

Perhaps the ways in which writing is both imaginatively fascinating and physically and mentally draining is best demonstrated in the case of Mary Erle, a failed painter gone writer in *The Story of a Modern Woman* (1894) by Ella Hepworth Dixon.[19] While writing lends Mary the potential for financial independence and meaningful employment, her health fails severely as a result of the strain she puts upon herself to create 'a little bit of real observation' (p. 147). The doctor warns her against the strains of working too hard, echoing late nineteenth-century medical fears about the dangers of women writing, and suggests she leave London and treat herself with arsenic and strychnine. Mary decides to buy the medicines, but her comment that '[w]e've got to be dosed with poisons to make us fit to sit at a desk and write' is a telling description of the ways in which writing acts as both a toxin and a cure (an intoxicating process) for women writers (p. 148).[20] Women writers'

---

[18] Both *The Beth Book* and Bram Stoker's *Dracula* (Oxford: Oxford University Press, 1996) were published in 1897, but although Grand could not have read Stoker's work before writing her own, there are striking resemblances between this passage and Stoker's descriptions of Lucy as she changes into a vampire: Lucy, who is naturally pale and 'bloodless' (Stoker, p. 111) as a vampire has lips 'crimson with fresh blood' and blood on her face (p. 211), which seems suggestive of Beth's own streaked red and white complexion.

[19] The name Mary Erle is probably an overt reference to Marion Erle in Elizabeth Barrett Browning's *Aurora Leigh* (1856). Like Dixon's Mary, Marion Erle never marries, but is also freed from having to be dependent on a man.

[20] While I discuss the perceived dangers of writing for women in Chapter 4, Jane Wood, in *Passion and Pathology in Victorian Fiction* (Oxford: Oxford University Press, 2001), particularly emphasizes medical concerns for the New Woman's intellectual endeavours. For discussions of the New Woman, health, and gender roles, see, for example, Patricia Murphy, 'Reevaluating Female "Inferiority": Sarah Grand versus Charles Darwin', *Victorian Literature and Culture*, 26 (1998), 221–36, and Angelique Richardson, 'Allopathic Pills? Health, Fitness and New Woman Fictions', *Women: A Cultural Review*, 10 (1999), 1–20. *The Story of a Modern Woman* is often read as a deeply pessimistic text, an example of the failure of women to achieve both financial independence and happiness. See, for example, Erin Williams, 'Female Celibacy in the Fiction of Gissing and Dixon:

paranormal insight only serves to exacerbate the oppositional potentialities of their artistic discipline: it gives them creative power and takes away their physical health, offers them literary greatness and withholds it by denying them the ability to fully recover from their moments of ecstatic inspiration.

Ultimately, Beth decides to become a feminist public speaker. Writing is indeed an initiatory process for her, a trial which she must undergo in order to prepare her for her career in oratory. As her friend Angelica suggests, '[y]our writing may have helped to perfect you in some other form of expression' (p. 520). Even as a public speaker, however, Beth inevitably returns to her writing, drafting manuscripts that she will recite. Although the process of speaking and writing are delicately intertwined, Beth is eager to dismiss writing as a distraction from her true calling in life: '[s]he had been misled herself, and so had everyone else, by her pretty talent for writing, her love of turning phrases, her play on the music of words. The writing had come of cultivation, but this – the last discovered power – was the natural gift' (p. 525).

This passage, however, undermines Beth's struggles with her writing and glosses over the great difficulties and triumphs she faced as a writer. Once again, this is suggestive about Beth's discomfort with her own extrasensory perception: when she claims that oration is her 'natural gift', her writing, by contrast, becomes an unnatural or supernatural gift, one that must be forsaken for its potential threat to Beth's political agenda. While she fears that her supernatural writing may have jeopardized her claims to speak seriously for the plight of women, and that oratory is the most natural and resolute means of expressing her views, her public speaking is shown to be influenced by the very kind of heightened power she is anxious about. When she speaks, she holds her audience under some kind of magical spell, hypnotizing and mesmerizing them with her words (she 'held them with curious talk' [p. 525]) and bringing them at last to 'a kind of awakening' (p. 419). Indeed, she is captivated by her own power. Ideala suggests that she will be 'impelled to choose' (p. 391) a means of expressing herself, and Beth herself believes that she is propelled into the career by an outside force: 'her vocation – discovered by accident, and with dismay for it was not what she would have chosen for herself in any way had it occurred to her that she had any choice in the matter' (p. 524). The choice of the word vocation is significant, for while it implies a career choice, it also connotes divine or spiritual influence.

The Silent Strike of the Suburbanites', *ELT*, 45 (2002), 259–79. Williams argues that for Mary, 'acceding to marriage comes at the cost of liberty, but rejecting it dooms one to bitter solitude and social marginality. With the past lost and the future inaccessible, the narrative forecloses upon any prospect of progress towards either felicity or independence' (p. 272). I suggest that *The Story of a Modern Woman* can be read as a more positive text, in which Mary both retains her liberty and supports herself with her writing. Margaret D. Stetz is also making strides towards reclaiming Dixon's oeuvre from such overt pessimism and suggests that Dixon often uses humour to discuss gender relations. See Margaret D. Stetz, 'The Laugh of the New Woman', in *The Victorian Comic Spirit: New Perspectives*, ed. by Jennifer A. Wagner-Lawlor (Aldershot: Ashgate, 2000), pp. 219–41.

Beth both chooses and is chosen, mesmerizes and is mesmerized by her speaking. Speaking is 'surpris[ing] and bewilder[ing]' (p. 419), an 'agony' (p. 420), but also a process which leaves her 'awestricken' (p. 420). Expression itself is exhilarating and terrifying for Beth, a moment of ecstasy which brings her into a trance state and fills her with both rapture and bewildered horror at making a public appearance.[21]

Grant Allen's *The Type-Writer Girl* and Powerful Automaticity in New Woman Writing

In Chapter 1 I examined the uncanny effect of technology on the nameless telegraph girl in James's 'In the Cage', whose social position is so helpless that she is ultimately denied a voice and is made automatic by her daily repetitive and mindless tasks. Indeed, while James and other Gothic writers such as Bram Stoker are concerned with technologies of writing and the New Woman, they figure women writers as being trapped into automaticity by the machine. For example, Mina in *Dracula* may keep a journal in which she records her private thoughts and anxieties, but Stoker suggests that her role in the novel is to act as a transcribing machine for the words of the other characters: she types out all of the journal entries, letters, telegrams, and written material pertinent to understanding and finding Dracula, and then copies it out.[22] She even transcribes Dracula's thoughts when hypnotized, repeating what he sees and hears in a way that reminds Harker 'as though she were interpreting something'.[23] Although the word 'interpret' suggests agency, an ability to decipher and to make intelligible, in this excerpt it suggests rather that Mina is simply reading aloud the thoughts of the Count, conveying the message without any adaptation or clarification of its contents. Despite the fact that, under hypnosis, Mina and Dracula share an exchange of powers and

---

[21]   Verena Tarrant in Henry James's *The Bostonians* also seems to go into a trance state while she speaks. In 'Possession and Personality: Spiritualism in *The Bostonians*', *American Literature: A Journal of Literary History, Criticism, and Bibliography*, 49 (1978), 580–91, Susan Wolstenholme argues that Verena's trance-like oration and her relationships with the other characters in the novel suggest that she 'is simply lost in vacuity [...]. She is lost amid the inner working of her own mind, which consists of nothing but the personalities of the people around her' (p. 591). Wolstenholme fails to recognize, however, the mesmeric effect Verena has, not only on her audiences, but also on Basil Ransom and Olive Chancellor. As I argue in Chapter 3, mesmerism is a dynamic site of the exchange of powers, in which the mesmerist and the mesmerized share control. Although Verena does not attain the same ecstatic heights of perception open to Beth, she nevertheless finds an element of power in her public speaking.

[22]   For discussions of *Dracula* and the New Woman see, for example, Marie Mulvey-Roberts, '*Dracula* and the Doctors: Bad Blood, Menstrual Taboo and the New Woman', in *Bram Stoker: History, Psychoanalysis and the Gothic*, ed. by William Hughes and Andrew Smith (Basingstoke: Macmillan, 1998), pp. 78–95.

[23]   Stoker, p. 312.

knowledge which leads to Dracula's destruction, Stoker denies Mina decisive authorial agency, and even uses her to mock the figure of the New Woman. On the subject of marriage, for example, Mina sarcastically remarks,

> [s]ome of the New Woman writers will someday start an idea that men and women should be allowed to see each other sleep before proposing or accepting. But I suppose the New Woman won't condescend in future to accept; she will do the proposing herself. And a nice job she will make of it too![24]

Both 'In the Cage' and *Dracula* use the supernatural (telepathy in 'In the Cage' and hypnotic control in *Dracula*) and concepts of uncanny technology to identify the writing female body with the writing machine, a process which ultimately jeopardizes their identity. New Woman fiction, however, suggests that women writers are more positively accessing the writing machines within themselves: automatic writing serves not to subsume women into the machine, but rather to invest them with the ability to use automaticity to tap into hidden springs of creativity and independence.

In the Victorian period, typewriting was associated not only with automatic writing, but also with the New Woman, who often chose to work as a typist to earn a living. New Woman fiction frequently depicted the figure of the woman typewriter. Rachel in *Red Pottage*, for example, works as a typewriter for several years before coming into her inheritance. In George Gissing's *The Odd Women* (1893), Mary Barfoot and Rhoda Nunn teach young women typewriting so that they can support themselves. While the fact that 'typewriter' could refer to a woman and/or a machine is suggestive about the working woman's reception in a society anxious about her entering the public sphere, this chapter argues that women have a more complicated relationship with technologies of writing, one which gives them an energy to create and critique (both the ills of society and their own writing) and to play both an active part and a passive role in the act of transcription.

Typewriters such as Juliet Appleton in Grant Allen's *The Type-Writer Girl*, for example, transcribe both analytically and mechanically, choosing key moments to either actively engage with or automatically type the words at their disposal.[25] For example, Juliet gives a close reading of the poetry she transcribes and deduces that the verses are about Mr. Blank's love for her. While Juliet is a writing machine, in this moment of analysis she also has agency: she criticizes Blank's poetry, and writes a short story and later a novel. Blank, on the contrary, is an empty page, his name implying that he is himself devoid of critical thought and, although an editor, not always in control of his writer. Indeed, Juliet takes on the masculine role in their relationship, projecting her ideals onto him and composing him as she would compose a letter.

---

[24]   Stoker, p. 89.

[25]   Grant Allen, *The Type-Writer Girl* (Peterborough: Broadview, 2004). Subsequent references to this edition are given after quotations in the text.

Juliet's experience using a typewriter suggests that even this most monotonous and mindless writing technology pushes women writers towards realizing literary potential. While this process is an initiatory one, in which Juliet suffers through the repetitive and dull nature of clerical work, poverty, and the realization that she belongs to a class of society which makes her an unsuitable match for Blank, she also seems to take theatrical delight in unhappy situations which she can later use in her novels. As her name implies, Juliet has a flair for drama, romance, and, of course, fiction, a trait that is emphasized when she re-names Blank as Romeo.

Indeed, the whole novel is a retelling of Juliet's unfortunate love affair, but it is also a story about transformation, about how she earns a living and develops into a writer. When she first begins work as a typewriter, Juliet writes dispiritedly, 'However, a type-writer I was, and a type-writer I must remain' (p. 28), and resents having to 'click, click, click, like a machine that I was' (p. 35). By the end of the story, however, she proclaims proudly to Michaela, '*I* am the type-writer girl!' (p. 139, emphasis in original), which suggests she embraces her identity, not only as a member of a professional class, as a woman in the workforce, and as the typewriter who stole Blank's heart, but also as the woman who, while yet a typewriter, refuses to become a passive female writing machine. Her transformation from being 'a' typewriter girl to 'the' typewriter girl is also significant, because it suggests a shift away from the anonymity of the masses of typewriter girls to the individual and independent woman she has become. She believes that '[n]o woman is born to be merely a type-writer', and indeed, she also becomes a novelist (p. 108). At the end of the novel, Juliet writes, '[i]f this book succeeds I mean to repay Michaela. Meanwhile, in any case, I am saving up daily every farthing to repay her. For I am still a type-writer girl – at another office' (p. 139). This self-referential, mischievous passage implies that the book does indeed succeed: the reader has access to the book because it has been published, and is in the reader's hands.

Significantly, Allen writes under the female pseudonym Olive Pratt Rayner in *The Type-Writer Girl*, a 'narrative cross-dressing' which, according to Clarissa J. Suranyi, 'initially creates the illusion of authenticity: who better to describe the experience of a typewriter girl than the girl herself? It also reveals, however, that the text is itself an act of impersonation or masquerade'.[26] What Suranyi seems to overlook, however, is the dynamic of automatic writing and mediumship, in which agency and authorship are often in flux. Allen's choice to write using a woman's name plays with notions of authorship, suggesting that he is perhaps himself a kind of medium or a writing machine for another voice; not the voice of the editor dictating to his typewriter, but the voice of the New Woman dictating at the fin de siècle.[27]

---

[26] Clarissa J. Suranyi, 'Introduction', in Allen, *The Type-Writer Girl*, pp. 9–17 (p. 10).

[27] For more on Allen's use of a female pseudonym and his decision to write New Woman fiction (contemporary critics wondered how and if a man could understand women enough to write New Woman fiction), see Vanessa Warne and Colette Colligan, 'The Man

Allen's depictions of women typists present a more hopeful vision of writing technology than either 'In the Cage' or *Dracula*, texts which offer claustrophobic visions of the meeting between woman and machine. An examination of New Woman fiction about women writers suggests that during the writing process, women find the threatening possibilities of becoming automatic conducive. Although Jill Galvan argues that '[a]utomatism did not just render women ignorant of what they transmitted: it converted them into unresisting bodies and thus all the more efficient in others' projects of diffusing (sending) or gathering (receiving) knowledge', I suggest that in New Woman fiction the unconscious, telepathic, and trance states women succumb to in order to incite the writing process may automate them, but this automation is also exhilarating, and these states give them heightened agency and perception.[28] The meeting between the machine and the body is both frightening and desirable, posing a risk to the autonomy of the self but also allowing women to attain powerful writing abilities.

In *The Beth Book* Beth is aware of both the threat and the fascination of becoming a machine. When she discusses the crowds of people in London, she believes that 'The friction of the crowd rubs out their individuality. In a crowd I feel mentally as if I were a maze of telegraph wires. The thoughts of so many people streaming out in all directions about me entangle and bewilder me' (p. 373). The word 'feels' in this passage suggests that Beth has intent: She is not mechanized by the telegraph wires, but instead has her own agency. She chooses to threaten the control she holds over herself in order to identify with the machine and harness its power. While here Beth may be anxious about her own agency and autonomy, concerned that her role as a writer may be that of the telegraph machine in which messages simply pass through her, her relationship with her own telepathic-like writing abilities is actually much more complicated. In this passage Beth becomes a medium, both the telegraph wire and the artist who walks across it. She becomes Nietzsche's vision of man as a tightrope walker: 'Man is a rope, fastened between animal and Superman – a rope over an abyss. A dangerous going-across, a dangerous wayfaring, a dangerous looking-back, a dangerous shuddering and staying-still.'[29] In embracing the liminal, and in becoming part of

---

Who Wrote a New Woman Novel: Grant Allen's *The Woman Who Did* and the Gendering of New Woman Authorship', *Victorian Literature and Culture*, 33 (2005), 21–46. Warne and Colligan argue that Allen's 'pseudonym suggests the complexities surrounding male authorship' (p. 43): Allen saw himself as 'a New Man prophet' who could predict and usher in the era of the New Woman (p. 27). However, I suggest that in order to speak for women, he had to speak as a woman, and rather than acting as the harbinger for New Woman ideas, he was a kind of medium for the political desires of the New Woman.

[28]   Jill Galvan, *The Sympathetic Medium: Feminine Channeling, the Occult, and Communication Technologies, 1859–1919* (Ithaca: Cornell University Press, 2010), p. 19.

[29]   Friedrich Nietzsche, *Thus Spoke Zarathustra*, trans. by R.J. Hollingdale (London: Penguin, 1969), p. 43. *Zarathustra* was published in parts between 1883 and 1885, and as a single volume in 1892.

the 'maze of telegraph wires', Beth develops heightened consciousness as well as a literary voice. Inspiration is an ecstasy for Beth, in the true sense of the word.

While I have already suggested that ecstasy is linked to a trance state, its etymology also connects it both to the uplifting of the soul and to insanity and bewilderment. When she writes, Beth is suspended between rapture and insanity, between dream and nightmare, and it is this moment of suspension that fascinates her. Beth is comparable here to Cosima, who is somehow psychically attuned to the world around her, suspended on the strings of the violin and delighting in being swept away by the vibrations of her own writing. Indeed, this passage implies that Beth also delights in balancing on the wires between her own will and the influence of the machine; she thrills in being in the state of suspension and oscillation. There is also a sense that she finds the notion of being swept away by the people in the London crowd intoxicating (much like Sherlock Holmes, who 'love[d] to lie in the very centre of five millions of people, with his filaments stretching out and running through them'), and that she revels in the possibility of being carried away by the masses.[30] Perhaps this desire is also one for her writing to be taken up by and spread through the crowd. Beth is fascinated by the potential for writing a bestseller which would terrifyingly, yet thrillingly, jeopardize her agency over her work and give her the acceptance and recognition she yearns for.

The Woman Writer in Paston's *A Writer of Books*: Masking the Message

In *A Writer of Books*, Cosima's first novel is significantly called *A 'Prentice Hand*, which conjures up images of a disembodied or severed hand, and also of the hand of the automatic writer at the séance, or the hand of the writer of the novel. The title brings up questions of agency: Whose hand wrote the book? Who inspired the writing? Who was the master to the apprentice?

Unlike Beth, Cosima is certain from the beginning that she wants to write. Significantly, she speaks of writing as her 'profession' (p. 23), which gives an insight, not only into editor Mr. Carlton's views on women writers, but also into how society viewed the New Woman writer:

> Carlton eyed her with some amusement. He knew plenty of girls who confessed to 'writing a little,' or 'scribbling nonsense,' but he had never before met one who announced that she had adopted literature for her profession with as much assurance as though she had said she was going to be a governess or a hospital nurse. Yet she did not look like a writing woman, he decided, for he shared the old-fashioned prejudice against literary ladies as a class, though when individually young and pretty he was prepared to forgive them everything, even success. (p. 24)

---

[30] Arthur Conan Doyle, 'The Adventure of the Cardboard Box', in *The Penguin Complete Sherlock Holmes*, preface by Christopher Morley (Harmondsworth: Penguin, 1981) , pp. 888–901 (p. 888).

Women writers, according to Carlton, should not have the confidence to assume writing as their profession, but rather should demonstrate stereotypical feminine hesitancy and modesty for their work: they should deny that they write to earn a living, and instead treat writing as a humble, even forgivable, pastime. Financial independence is something that is achievable for the male writer, but should not be considered for the woman writer. This passage suggests that while male writers can choose professional writing as a career, women writers must only hope for modest recognition, and should therefore turn to other, generically female occupations such as nursing or teaching positions. Writing professionally is somehow distasteful in a woman, not only because professionalism and intellectualism are considered masculine traits in the late nineteenth century, but also because of the notion that women writers are physically less feminine. The writing woman is thought to be unattractive, frumpy, and singularly unfeminine, or, as Dr. Maclure describes Beth in *The Beth Book*, '[c]oarse and masculine' (p. 366).

The associations between literary women and masculinity meant that women writers were uncomfortably juxtaposed between the masculine/public and feminine/private spheres. Journals, articles, and cartoons of the 1890s were full of descriptions, criticisms, and satires of the New Woman writer, many of them negatively portraying women writers' talents and physical appearance.[31] Ann Heilmann discusses the confliction women writers felt about their literary success: 'feminist writers problematize the conflicting desires and pressures women artists feel when their private and public roles are in collision. In particular, they explore the precarious balancing act women artists have to perform between conforming to traditional notions of feminine morality and securing their individual professional survival'.[32] Women writers faced considerable difficulties in a literary and social market that was hostile to women's public success as well as to their attempts to gain financial independence.

Indeed, as Heilmann suggests, the woman writer's 'private and public roles are in collision', but this collision does not mark simply a problematic. Instead, this section will suggest that gaining financial independence is itself uncanny for women writers of the fin de siècle, and that the act of writing, which takes women writers like Cosima to the very heights of perception, also allows them to cross the boundaries between traditional ideas about masculinity and femininity. Cosima is shrewdly able to disguise her fiction so that it pleases the mass market. At the same time, she addresses issues of realism and the plight of women, which she

---

[31]   See, for example, Sally Ledger and Roger Luckhurst, *The Fin de Siècle: A Reader in Cultural History, c. 1880–1900* (Oxford: Oxford University Press, 2000). Of course, depictions of the New Woman were not uniformly critical. Constance Harsh's article 'Reviewing New Woman Fiction in the Daily Press: *The Times*, the *Scotsman*, and the *Daily Telegraph*', *Victorian Periodicals Review*, 34 (2001), pp. 79–96, suggests that reviews of New Woman fiction were not always negative and that 'the late-Victorian critical practice was a complicated and heterogeneous affair', implying that while the New Woman writer was a controversial figure, her fiction did receive some acclaim (p. 91).

[32]   Heilmann, *New Woman Fiction*, p. 159.

feels are intrinsic to her literary ideals. Her training as an author allows her to inhabit different social roles and to convince her editor that her work conforms to his standards, a clever pretense which gives her absolute authority and agency over her work. The fact that she describes her work as a 'profession' is striking, not only because it implies she is entering a masculine sphere of work, but also because the word itself suggests deceit and artifice: after all, a profession can be defined as a declaration which can be either true or false. Cosima 'professes' to her editors and to her public that her work is the conventional triple-decker with a happy ending, but she is actually producing writing which conveys her realist and feminist message, and which places her at the intersection between her private passions and the reinventions of those passions she creates for her editor.

While Cosima is able to negotiate between her literary ambition and the requirements of her editor, she is still anxious about her role as a woman writer, which threatens to overthrow the balance between her public and private life. When she first considers marrying Tom, she believes:

> She would so enjoy proving in her own person the fallacy of the prejudice against the domestic capabilities of literary ladies. Like the heroine of an old-fashioned novel, she would arrange that her household should be a model of order; she would attend personally to her husband's comforts, give him nice dinners, darn his socks and sew on his buttons with her own hands. He would never see her with inky fingers or disheveled hair, but when he came home tired in the evening he should always find her prettily dressed and cheerful and good–humoured [...]. It was quite a charming little picture, and – the lonely old age of the needy spinster was not a pleasant fate to look forward to. (p. 112)

Here Cosima is anxious about becoming the stereotypical literary lady who lacks domestic happiness. Significantly, the fact that she imagines herself as a heroine in an old-fashioned novel suggests how unrealistic this picture of domestic bliss really is: a fiction, a romance, a fairy tale. But the fact that Cosima, whose literary ambition is 'to learn something of all sides of life and all sorts of conditions of men' (p. 45) and to write a truly new kind of fiction that describes the condition of women in the 1890s, imaginatively sees herself as a conventional heroine in a conventional plot suggests how uneasy she feels about her role as both a woman and a woman writer in a society hostile to women's ambition.

Cosima's marriage to Tom is disastrous because she discovers that in marrying him, she must sacrifice most of her independence: she has become more of a wife than a writer. The balance between the public and the private has been overthrown, and Cosima is left to struggle with the feelings of boredom and futility of the housewife: 'her once fertile brain seemed to have become dull and barren, her imagination absolutely refused to work, and she began to fear that her marriage, so far from exercising a stimulating influence upon her mind, had deadened or destroyed her literary faculty' (p. 142). That Cosima's imagination can be both 'fertile' and 'barren' in a passage about the unhappiness she finds in her marriage seems to suggest that she is attempting to cleanse the words of their association with child-bearing and the child-rearing duty she is expected to fulfil as a woman and a wife. The desire to produce a work of fiction, to cultivate her writing and for it to flourish, has replaced the maternal instinct to 'produce' children. The word 'faculty' here is also suggestive. Paston suggests that Cosima's writing is aided by a 'sixth sense' or an intuitive faculty (p. 232). Cosima's marriage seems to temporarily strip her of this power, and it is only when she finds the courage and resources to leave Tom that she fully regains it. Significantly, Cosima's decision to end her marriage not only gives her access again to the moments of supernatural inspiration which fuel her work, it also allows her to find a balance between conforming to society's expectations and breaking them: Cosima discovers that she can be a writer *and* embody feminine traits, she can enter the public sphere as a woman writer and retain her female identity without trepidation. When Cosima discovers Tom's infidelity, she realizes that she has the right and the moral responsibility to herself to find the happiness she never found as his wife.

Cosima also argues that women have a greater purpose than simply to follow the traditional paths of finding love and then marrying:

> [l]ove may once have been a woman's whole existence, but that was when a skein of embroidery silk was the only other string in her bow. In the life of a modern woman, blessed with an almost inexhaustible supply of strings, love is no less episodical than in the life of a man. (p. 257)

The 'strings' in this passage allude once again to Cosima as a violinist, whose 'imagination had now become a well-tuned instrument upon which she played with a virtuoso's hand' (p. 206). Comparing Cosima to a violin whose strings attune her to supernatural perception again links her to the discourse on mesmerism. Phyllis Weliver quotes from Mesmer's *Dissertatio physico-medica de planetarum influxu* (1766) to demonstrate how Mesmer linked music, and in particular the violin, to mesmerism:

> The harmony established between the astral plane and the human plane ought to be admired as much as the ineffable effect of UNIVERSAL GRAVITATION by which our bodies are harmonized [...] as a musical instrument furnished with several strings, the exact tone resonates which is in unison with a given tone. Likewise, human bodies react to stellar configurations with which they are joined by a given harmony.[33]

In accessing the trance state of mesmerism, Cosima finds harmony with her identity as a woman writer, a harmony she lacks in her dealings with the discordant elements of the literary and marital marketplace.

Not only does Cosima write mesmerically, her thoughts like strings (or Beth's telegraph wires) which vibrate with inspiration from a power outside of herself and

---

[33] Weliver, pp. 63–4.

which she harnesses for use in her novels, but there is also a suggestion that this kind of writing enables her to find the 'harmony of the spheres', a balance between the public and private spheres.[34] She has many strings in her bow, many possibilities in both her career and private life, which do not restrict her to love and marriage alone. When Cosima suggests that life and love are 'episodical' (p. 257), she implies that these episodes are not only a series of incidents, or the episodes of an illness or psychological state (like the trance-like state of awareness in which she mesmerically writes), or even episodes as musical forms, which reinforces her connection to music, the figure of the violinist, and mesmerism, but also the fictional narratives which make up a story. Love itself is a fiction, and women like Cosima can now focus their energies, not on sustaining a conventional lifestyle, but on career and success.

Cosima suggests here that the woman's sphere is changing and expanding, and that authorship itself gives women an enriching life experience. In negotiating between her public, her editor, and her own ideals, Cosima comes to symbolize the New Woman writer herself, whose ability to disguise herself and adopt different social roles makes her a heightened figure, both a woman and a New Woman indeed, with desires for career, independence, and success. However, while the possibility of being published and achieving success holds a fascination for women writers like Cosima, it is one which is both attractive and repellent.

Indeed, her name may allude to Cosima Wagner (1837–1930), since both Cosimas had links to music and had unconventional lifestyles. Cosima Wagner had an affair and several children with the composer Richard Wagner before marrying him, but she also achieved fame and recognition in her own right: after her husband's death in 1883, she became the director of the Bayreuth Festival, and her diaries recording her life with Wagner have since been published.[35] Cosima's name suggests that like Cosima Wagner, she may be famously scandalous because of her decisions to disregard marital conventions, but also that she will have a 'cosmic' impact with her writing: potentially, her work will gain universal acceptance.

Cosima admits that she wants to write 'for name and fame' (p. 210), but she makes a clear distinction between literary and popular success, and while she

---

[34]   In a section entitled 'Separate Spheres' in *Women Musicians in Victorian Fiction* (pp. 38–47), Weliver discusses the problems women faced in trying to find a balance between the career and the home: 'far from allowing their marriages to exclude careers, or vice versa, professional female musicians from all classes began to have both, despite the fact that combining career and marriage raised complex questions about a woman's place and how she might reconcile the independence required in professional life with the domestic role expected of ideal Victorian wives' (p. 38). See also Sarah A. Wilburn's *Possessed Victorians: Extra Spheres in Nineteenth-Century Mystical Writings* (Aldershot: Ashgate, 2006), which suggests that women could inhabit 'extra spheres' through being possessed in a spiritualist séance: women could have access to public, private, and mystical spheres in Victorian society.

[35]   See Cosima Wagner, *Cosima Wagner's Diaries*, ed. and annotated by Martin Gregor-Dellin and Dietrich Mack, trans. and intro. by Geoffrey Skelton (London: Pimlico, 1974).

refuses to write a bestseller, she desires from a young age to gain celebrity. For example, she keeps a journal, 'the raw material of the masterpiece which, she had already decided, should make her famous in the future' (p. 6). The literary marketplace is both a venue for gaining literary celebrity and, what Cosima dreads, where her writings could be misinterpreted or misunderstood. Just as Cosima fears that elements of the supernatural in her writing will damage her aim of literary realism, celebrity itself, the possibility of being recognized for her ability to create a 'masterpiece' (p. 6), also poses a threat to her political agenda. While fame for her is desirable, achieving fame might disturbingly imply that she has conformed to male literary expectations of great writing, instead of adhering to her own feminist and realist ambitions. The elusiveness of celebrity for Cosima also makes it a ghostly presence in the text, and she is haunted not only by the possibility that she will achieve fame as a novelist, but also by the possibility that she might not. Attempting to challenge conventional expectations is problematic for women writers, as Heilmann suggests, because it forces them to find a way in which they can maintain their feminine morality as well as gain literary success. Ultimately, however, it brings them into contact with the male publishing world, where they learn to powerfully and authoritatively negotiate between their own literary ambitions and the requirements of their editor and their public.

Cosima's authorship, as suggested earlier, is influenced by supernatural perception and inspiration, but she is also influenced by male expectations of what should be sold in the literary marketplace. While this is not to suggest that a male-dominated publishing world has a supernatural power over Cosima's literary output, it does suggest that for Cosima to achieve literary recognition and financial independence in a male-dominated profession is uncanny, and that in successfully entering a traditionally masculine sphere of work she is powerfully and persuasively balancing her own ambitions with those of her editor. Cosima's writing is suggestive of the ways that New Woman writers had to be adept at disguise and camouflage, superficially pleasing their editors and the reading public while still communicating their observations on real life and the plight of women at the end of the century.

The 1890s saw a rise in shorter volumes, which were more affordable than the triple-deckers which had previously dominated the market and lessened the stronghold of lending libraries like Mudie's and Smith's. Although earlier in the Victorian period, 'Mudie and Smith could ruin an author's career by refusing to carry his or her novels', writers at the fin de siècle were still confronted with the problem of having to censor their work to suit both the reading public and the conventions of the publishing world.[36] Women writers had the added pressures of being judged by male editors and seeing their work stripped or banished by male censorship: women had to struggle with the fact that their writing was filtered through a masculine gaze before it ever reached the reading public.

---

[36]   Steve Farmer, 'Appendix F: Literary Censorship in Victorian England', in *The Story of a Modern Woman*, pp. 267–88 (p. 267).

The reaction of many of the male characters to Cosima's writing suggests the ways in which women writers faced both hostility and scepticism in the literary marketplace. While the editor, Mr. Carlton, does not take her sense of professionalism seriously, he also believes that women can only write well once they have faced great hardships: 'Then, when you were faded, and lonely, and disillusioned, and middle-aged, you might write a great novel' (p. 55). Although Carlton fails to recognize the triumphant outcome which follows this trying period, he nevertheless suggests that writing is an initiatory process. Tom's attitude towards Cosima's writing is practical but unsympathetic: he believes she should sacrifice her talent in favour of money. Mr. Haddon of Haddon & Waller will only publish her novel once she has given it a 'happy ending[s]' that will please the public (p. 79). Cosima's struggles with male editors are similar to those of Mary Erle in *The Story of a Modern Woman*, whose novel must be censored because she has written about 'a young man making love to his friend's wife' (p. 146). While Mary defends herself for her 'real observation' (p. 147), the editor dismisses her, arguing that '[t]he public won't stand it, my dear girl. They want thoroughly healthy reading […] Must be fit to go into every parsonage in England. Remember that you write chiefly for healthy English homes' (p. 146). Cosima's and Mary's editors both suggest that what the reading public wants is what the male public wants: not stories about the plight of women, about marital affairs, or, as in Cosima's story, about an unhappy couple, but conventional stories with conventional endings.

In order to succeed, however, Cosima does not simply comply with her editor's advice or 'sell out' to the popular market. Instead, she skilfully disguises her writing in order to superficially please her editor, simultaneously keeping her real observations veiled within her work. In this way, her work maintains its edginess as well as her commitment to a political agenda. In writing her first novel, Cosima 'was forced to remind herself that if she were to ever gain a hearing, she must begin by suiting herself to the requirements of the public, or at any rate of the publishers' (p. 81). She edits her work, seemingly to the standards that Mr. Haddon requires: she 'transformed tragedy and tears into the conventional white satin, wedding-bells and prospective happiness' (p. 81). Her transformation of her novel implies, however, that not only does she change the work to seemingly meet Haddon's expectations, but she also transforms it into a more polished, precise piece. She maintains that she cuts out the most 'effective' (p. 81) parts of her writing, but admits they are also the most 'redundant' (p. 81); the work she submits may seem to be written to Haddon's specifications, but Cosima is actually honing her skills and making more precise her ideas and characters.

Cosima's second 'book sold more freely [than the first], and there seemed a chance that in this case the author's profits might actually amount to two figures!', which suggests that she has become even more adept at disguising her own ideas within her work (p. 143). Indeed, the only character who is aware of the artifice is Quentin Mallory, who says that 'the machinery [in the novel] is too apparent […]. And the author makes gallant, though happily not always successful, efforts to appeal to the average reader' (p. 143). Although Quentin claims to see through

Cosima's charade, the public and her editor are fooled: the book is well received and sells well. Furthermore, although he knows that her writing panders to the public, Quentin's recognition of the hidden message in her book suggests that her political message is still a substantial, if submerged, part of her work.

Cosima's ability to disguise herself so well depends on her ability to brutally edit her work, a process which is comparable to completing a ritualistic slaughter. Cosima transforms into the symbolic figure of a priestess who, in a moment of controlled violence, effectively edits and improves (or redeems) her work:

> [d]uring the next few days Cosima was occupied in slashing and mutilating the offspring of her brain, and soon began to feel as if she were up to her elbows in gore. As she sacrificed her most effective, but at the same time redundant passages, she compared herself to the Russian mother who flung some of her children out of a sledge to the wolves in order to save the rest. (p. 81)

In her editorial amputations, she is successfully cutting away the less effective moments in her writing, a process which is empowering because it strengthens her work. Like the Russian mother who sacrifices some of her children for the salvation of the others, Cosima must also be professionally ruthless in order to deliver her work from its weaker points, and also to enhance and enrich it.

The notion of Cosima cutting away the 'offspring of her brain' suggests that she is cutting out the part of herself that would link her to hysteria: she is symbolically cutting away her womb ('hysteric' is from the Greek word for womb) and stopping her menstrual cycle (she is up to her elbows in gore, perhaps symbolically stemming the flow of blood). This is not an attempt to destroy her identity as a woman, but rather an amputation of the negative connotations of hysteria which link the female body to female illness. Indeed, Cosima is embracing other aspects of hysteria, such as moments of extreme (or perhaps dangerous) excitement, which she can use to drive and inspire her writing.

For too long, New Woman novels have been read as stories of women's failures. Although women writers in New Woman fiction face many hardships, they also act as priestesses for the numinous, ritually sacrificing themselves at the altar of inspiration so that they can resurrect themselves as powerfully transformed figures. New Woman writers may be anxious about engaging with the supernatural in their novels, but it is the thrilling and destructive nature of extrasensory perception which makes the figure of the writing woman a force to be reckoned with at the end of the century.

Postscript

The Late Victorian Gothic: Mental Science, the Uncanny, and Scenes of Writing has explored how writers and mental scientists of the fin de siècle were facing a new kind of Gothicism in which they were radically conflicted between a desire to police the boundaries of science, identity, and writing itself and, conversely, to experience the ecstasy of engaging with the supernatural. Although D.H. Lawrence argues that 'the Wondrous Victorian Age managed to fasten the door so tight, and light up the compound so brilliantly with electric light, that [...] [t]he Unknown became a joke', problematically, Victorians in the 1880s and 1890s found that they could not, after all, fasten the door of empirical science against supernatural elements.[1] Indeed, even the electric light which 'brilliantly' lit up the compound was linked to the Unknown. Like other technological innovations of the period, electricity was seen both as a symbol of materialist science and as evidence that science could not escape connection with occult phenomena: science was haunted by the possibility that the very techniques it used to eliminate supernatural potential actually evoked it.

Late Victorians tried to keep out the Unknown, but the joke was not only on Lawrence (who failed to see the significance of Victorian supernaturalism on cultural, literary, and psychological developments in the twentieth century), but also on the late Victorians themselves. Indeed, they were unable to regulate the supernatural, but this was in part because they delighted in the haunted and occult elements which crept into fiction and mental science texts. Writers wanted to be ghosted by writing, desiring to give themselves up to moments of 'mysterious life-suggestion'.[2]

This book has treated haunted scenes of writing as crucial to an understanding of authorship, identity, interactions with new technologies, the science of mind, and spiritualism at the fin de siècle, which in turn raises questions about the fascination at the end of the century with resurrecting ghosts. Henry James conjured up the ghost of a pen on his death-bed (perhaps symbolically bringing back to life the James who wrote by hand rather than the James who dictated to his typist), and spirits were called from the grave by their relatives so that they could appear in photographs. Mesmerism revived all the occult associations in hypnotism, and Vernon Lee revived and exorcised ghosts in her aesthetic writings.

Why were writers in this period drawn to resurrecting ghosts in their work? Women writers were particularly concerned with symbolically bringing themselves back to life through their writing: in exploring their self-identification

---

[1]   D.H. Lawrence, *Kangaroo*, ed. by Bruce Steele, intro. by Macdonald Daly (London: Penguin, 1997), p. 285.

[2]   Ibid.

with ghosts and heightened perception, they could sacrifice themselves at the altar of inspiration and rise up again, phoenix-like. For women writers, ghosts became emblematic, not only of women's resistance against legal and political invisibility, but also for their numinous revival from patriarchal subjugation.

Perhaps the notion that writers obsessively resurrected and exorcised ghosts also has something to do with the term 'ghost' itself. One of the definitions for the word is '[u]sed as the conventional equivalent for L. *spiritus*, in contexts where the sense is *breath* or *a blast*'. A ghost is an afflatus and an inspiration which breathes life and creativity into authors, but 'ghost' can also be defined doubly as '[a] good spirit' and '[a]n evil spirit'. Writers at the end of the century were concerned about sites of power and influence, and ghosts may have offered a means of articulating how authorship itself was a balancing act between good and evil, an engagement with both the rapture and the terror of suspended agency.

Indeed, in writing about ghosts, did late nineteenth-century writers hope that this was a means of ensuring they would haunt the canon? Or does writing about the ghostly make authors spectral figures in literary criticism? I have attempted to resurrect some ghostly figures in the canon, and to show how even canonical writers are drawn to the supernatural. Paradoxically, the fact that they wrote about ghosts makes these authors the most material witnesses to fin de siècle anxieties and preoccupations about identity.

Bibliography

Primary Sources

Allen, Grant [1897], *The Type-Writer Girl* (Peterborough: Broadview, 2004).

Ambrose, Gordon and George Newbold [1956], *A Handbook of Medical Hypnosis*, 4th edn (London: Baillière, 1980).

Aster [1899], *The Bridge of Light: A Message from the Unseen* (London: Gay and Bird, 1899).

Aurea, Lux [1901], *Light from the Summerland* (London: Gay and Bird, 1901).

Baldwin, Louisa, [1895], 'Many Waters Cannot Quench Love', in *The Shadow on the Blind* (Ashcroft: Ash-Tree, 2001), pp. 59–67.

Barter, John [1890], *How to Hypnotise: Including the Whole Art of Mesmerism etc.* (London: Simpkin, 1890).

Besant, Annie [1892], 'Foreword' in *Nightmare Tales*, by H.P. Blavatsky, p. 3.

Blavatsky, H.P. [1892], 'A Bewitched Life (As Narrated by a Quill Pen)', in *Nightmare Tales* (London, 1892), pp. 7–67.

———— [1892], 'The Cave of the Echoes', in *Nightmare Tales*, pp. 68–80.

———— [1892], 'The Ensouled Violin', in *Nightmare Tales*, pp. 98–133.

Bourru, Hippolyte and P. Burot [1885], 'Un cas de la Multiplicité des états de Conscience Chez un Hystéro-Epileptique', *Revue Philosophique*, 20 (1885), 411–16.

———— [1888], *Variations de la Personnalité* (Paris: Baillière, 1888)

Braid, James [1843], *Neurypnology: or the Rationale of Nervous Sleep Considered in Relation to Animal Magnetism or Mesmerism*, ed. and intro. by Arthur Edward Waite, new edn (London: Redway, 1899).

Brooke, Emma Frances [1894], *A Superfluous Woman* (London: 1894).

Carpenter, William Benjamin [1874], *Principles of Mental Physiology* (London: Routledge, 1993).

———— [1877], *Mesmerism, Spiritualism, etc.: Historically and Scientifically Considered: Being Two Lectures Delivered at the London Institution* (London: Longmans, Green, 1877).

Charcot, Jean-Martin, 'Hypnotism in the Hysterical', in *A Dictionary for Psychological Medicine*, ed. by D. Hack Tuke, I, pp. 606–10.

Cholmondeley, Mary [1899], *Red Pottage* (London: Virago, 1985).

Coates, James [1897], *Human Magnetism or How to Hypnotise: A Practical Handbook for Students of Mesmerism*, new rev. edn (London: Nichols, 1904).

Coates, Mrs. James [1911], *Photographing the Invisible* (London: L.N. Fowler and Co., 1911).

Cobban, J. Maclaren [1890], *Master of His Fate*, ed. by Brian Stableford (Elstree: Greenhill, 1987).

Collins, Wilkie [1868], *The Moonstone* (New York: Knopf, 1992).

Corelli, Marie [1897], *Ziska: The Problem of a Wicked Soul* (London: Bristol, 1897).

Crookes, William [1872], *Correspondence upon Dr. Carpenter's Asserted Refutation of Mr. Crookes's Experimental Proof of the Existence of a Hitherto Undetected Force* (London: *Quarterly Journal of Science*, 1872).

——— [1874], 'Notes of an Enquiry into the Phenomena Called Spiritual', in *Researches in the Phenomena of Spiritualism* (London: J. Burns, 1874), pp. 81–102.

Davey, William [1854], *The Illustrated Practical Mesmerist: Curative and Scientific*, 2nd edn (Edinburgh, 1856).

Dilke, Emilia [1886], 'The Black Veil', in *The Shrine of Death and Other Stories* (London: George Routledge, 1886), pp. 79–84.

——— [1886], 'The Physician's Wife', in *Shrine of Death*, pp. 40–56.

——— [1886], 'The Shrine of Death', in *Shrine of Death*, pp. 11–24.

——— [1886], 'The Silver Cage', in *Shrine of Death*, pp. 27–36.

——— [1891], *The Shrine of Love and Other Stories* (London: George Routledge, 1891).

Dixon, Ella Hepworth [1894], *The Story of A Modern Woman* (Peterborough: Broadview, 2004).

Donkin, H.B. [1892], 'Hysteria', in *A Dictionary of Psychological Medicine*, ed. by D. Hack Tuke, I, pp. 618–27.

Dowie, Menié Muriel [1895], *Gallia*, ed. by Helen Small (London: Dent, 1995).

Doyle, Arthur Conan [1881], 'After Cormorants with a Camera', in *The Unknown Conan Doyle: Essays on Photography*, ed. by John Michael Gibson and Richard Lancelyn Green (London: Secker & Warburg, 1982), pp. 1–12.

——— [1881], 'To the Waterford Coast and Along It', in *Unknown Conan Doyle*, pp. 51–9.

——— [1882], 'On The Slave Coast with a Camera', in *Unknown Conan Doyle*, pp. 13–22.

——— [1883], 'A Few Technical Hints', in *Unknown Conan Doyle*, pp. 38–9.

——— [1883], 'Trial of Burton's Emulsion Process', in *Unknown Conan Doyle*, pp. 40–42.

——— [1887], *A Study in Scarlet*, in *Complete Sherlock Holmes*, pp. 15–88.

——— [1890], *The Sign of Four*, in *Complete Sherlock Holmes*, pp. 89–158.

——— [1891], 'The Boscombe Valley Mystery', in *Complete Sherlock Holmes*, pp. 202–17.

——— [1891], 'The Man with the Twisted Lip', in *Complete Sherlock Holmes*, pp. 229–44.

——— [1891], 'The Red-Headed League', in *Complete Sherlock Holmes*, pp.176–90.

——— [1891], 'A Scandal in Bohemia', in *Complete Sherlock Holmes*, pp. 161–75.

——— [1892], 'The Adventure of the Copper Beeches', in *Complete Sherlock Holmes*, pp. 316–32.

———— [1892], 'The Adventure of the Speckled Band', in *Complete Sherlock Holmes*, pp. 257–73.

———— [1893], 'The Adventure of the Cardboard Box', in *Complete Sherlock Holmes*, pp. 888–901.

———— [1893], 'The Crooked Man', in *Complete Sherlock Holmes*, pp. 411–22.

———— [1893], 'The Final Problem, in *Complete Sherlock Holmes*, pp. 469–80.

———— [1893], 'The Naval Treaty', in *Complete Sherlock Holmes*, pp. 447–69.

———— [1894], 'The Parasite', in *The Edinburgh Stories of Arthur Conan Doyle* (Edinburgh: Edinburgh University Student Publications Board, 1981), pp. 41–80.

———— [1901], *The Hound of the Baskervilles*, in *Complete Sherlock Holmes*, pp. 667–766.

———— [1911], 'The Adventure of the Red Circle', in *Complete Sherlock Holmes*, pp. 901–13.

———— [1919], *The Vital Message* (New York: George H. Doran, 1919).

———— [1922], *The Case for Spirit Photography* (London: Hutchinson, 1923).

———— [1922], *The Coming of the Fairies* (New York: Weiser, 1979).

———— [1923], 'The Adventure of the Creeping Man', in *Complete Sherlock Holmes*, pp. 1070–83.

———— [1924], 'The Adventure of the Illustrious Client', in *Complete Sherlock Holmes*, pp. 984–99.

———— [1924], 'The Adventure of the Sussex Vampire', in *Complete Sherlock Holmes*, pp. 1033–44.

———— [1926], *The History of Spiritualism*, 2 vols (London: Cassell, 1926).

———— [1927], 'The Adventure of The Veiled Lodger', in *Complete Sherlock Holmes*, pp. 1095–1102.

————, *The Penguin Complete Sherlock Holmes*, preface by Christopher Morley (Harmondsworth: Penguin, 1981).

Du Maurier, George [1894], *Trilby* (London: Everyman, 1994).

Galbraith, Lettice [1893], 'The Ghost in the Chair', in *New Ghost Stories* (London, 1893), pp. 51–66.

Galton, Francis [1878], 'Composite Portraits', *Journal of the Anthropological Institute*, 8 (1878), 132–44.

———— [1879], *Generic Images* (London: William Clowes and Son, 1879).

———— [1892], *Finger Prints* (London: Macmillan, 1892).

Gilman, Charlotte Perkins [1892], 'The Yellow Wallpaper', in *Daughters of Decadence: Women Writers of the Fin-de-Siècle*, ed. by Elaine Showalter (London: Virago 1993), pp. 98–117.

Gissing, George [1893], *The Odd Women* (Oxford: Oxford University Press, 2008).

Goodrich-Freer, A. [1899], ed., *The Alleged Haunting of B— House, Including a Journal Kept During the Tenancy of Colonel Lemesurier Taylor* (London: Redway, 1899).

———— [1899], *Essays in Psychical Research* (London: Redway, 1899).

———— [1900], ed., *The Professional: and Other Psychic Stories* (London: Hurst and Blackett, 1900).

Grand, Sarah [1893], *The Heavenly Twins* (London: Heinemann, 1893).

———— [1897], *The Beth Book* (London: Virago, 1980).

Gurney, Edmund [1880], *The Power of Sound* (London: Smith, Elder, 1880).

Gurney, Edmund, F.W.H. Myers, and Frank Podmore [1886], *Phantasms of the Living*, 2 vols (London: Trübner, 1886).

Haggard, H. Rider [1887], *She: A History of Adventure* (London: Longmans, 1887).

Hardy, Thomas [1874], *Far From the Madding Crowd*, ed. by Robert C. Schweik (London: Norton, 1986).

———— [1893], 'The Fiddler of the Reels', in *The Short Stories of Thomas Hardy* (London: Macmillan, 1928), pp. 401–19.

Hawthorne, Nathaniel [1852], *The Blithedale Romance* (London: Penguin, 1983).

Huysmans, Joris-Karl [1884], *Against Nature*, trans. by Robert Baldick (London: Penguin, 1959).

James, Henry [1886], *The Bostonians* (Harmondsworth: Penguin, 1984).

———— [1892], 'The Private Life', in *Complete Tales*, VIII, pp. 189–227.

———— [1898], 'In the Cage', in *Complete Tales*, X, pp. 139–242.

———— [1908], 'The Portrait of a Lady', in *Literary Criticism: French Writers; Other European Writers; The Prefaces to the New York Edition*, ed. by Leon Edel (New York: Library of America, 1984), pp. 1070–85.

————, *The House of Fiction: Essays on the Novel*, ed. by Leon Edel (London: Hart-Davis, 1957).

————, *The Complete Tales of Henry James*, ed. by Leon Edel, 10 vols (London: Hart-Davis, 1962–1964).

————, *Henry James Letters*, ed. by Leon Edel, 4 vols (London: Macmillan, 1974–1984).

————, *The Complete Notebooks of Henry James*, ed. by Leon Edel and Lyall H. Powers (Oxford: Oxford University Press, 1987).

James, William [1901], 'Frederic Myers's Service to Psychology', in *A William James Reader*, ed. by Gay Wilson Allen (Boston: Houghton Mifflin, 1971), pp. 155–64.

————, *Essays in Psychical Research*, ed. by Frederick Burkhardt, *The Works of William James*, 17 vols (Cambridge, MA: Harvard University Press, 1986), vol. XVI.

Janet, Pierre [1925], *Psychological Healing: A Historical and Clinical Study*, trans. by Eden and Cedar Paul, 2 vols (London: Allen & Unwin, 1925).

Journal of the Society for Psychical Research, 7 (1895–1896).

Keats, John [1819], 'This Living Hand', in *Complete Poems: John Keats*, ed. by Jack Stillinger (London: Belknap Press, 1982).

———— [1820], 'The Eve of St. Agnes', in *Complete Poems*, pp. 229–39.

Kellogg, John Harvey [1882], *Ladies Guide in Health and Disease, Girlhood, Maidenhood, Wifehood, Motherhood* (London, 1890).

King, John H. [1893], *Man an Organic Community: Being an Exposition of the Law that the Human Personality in all its Phases in Evolution, Both Co-ordinate and Disconsolate, is the Multiple of Many Sub-Personalities*, 2 vols (London, 1893).

Kingsford, Anna Bonus [1888], *Dreams and Dream-Stories*, ed. by Edward Maitland (London: Redway, 1904).

———— [1888], 'Steepside: A Ghost Story', in *Dreams and Stories*, pp. 116–46.

Kipling, Rudyard [1902], 'Wireless', in *The Best Short Stories* (Ware: Wordsworth Classics, 1997), pp. 143–58.

Külpe, Oswald [1895], *Outlines of Psychology: Based Upon the Results of Experimental Investigation*, trans. by Edward Bradford Titchener (London: Sonnenschein, 1895).

Lawrence, D.H. [1923], *Kangaroo*, ed. by Bruce Steele, intro. by Macdonald Daly (London: Penguin, 1997).

Lee, Vernon [1880], 'Faustus and Helena: Notes on the Supernatural in Art', in *Belcaro*, pp. 70–105.

———— [1880], *Studies of the Eighteenth Century in Italy* (London, 1880).

———— [1881], *Belcaro: Being an Essay on Sundry Aesthetical Questions* (London, 1881).

———— [1881], 'Chapelmaster Kreisler: A Study of Musical Romanticists', in *Belcaro*, pp. 106–28.

———— [1881], 'Cherubino: A Psychological Art Fancy', in *Belcaro*, pp. 129–55.

———— [1881], 'The Child in the Vatican', in *Belcaro*, pp. 17–48.

———— [1881], 'A Culture Ghost: or, Winthrop's Adventure', in *Hauntings: The Supernatural Stories*, pp. 251–76.

———— [1881], 'Ruskinism: The Would-Be Study of Conscience', in *Belcaro*, pp. 198–229.

———— [1882], 'Impersonality and Evolution in Music', *Contemporary Review*, 42 (1882), 840–58.

———— [1886], 'The Hidden Door', in *Hauntings: The Supernatural Stories*, pp. 321–38.

———— [1886], 'Oke of Okehurst; or, The Phantom Lover', in *Hauntings: The Supernatural Stories*, pp. 51–86.

———— [1887], 'Amour Dure' in *Hauntings: The Supernatural Stories*, pp. 7–30.

———— [1887], *Juvenilia: Being a Second Series of Essays on Sundry Aesthetical Questions* (London:. T.F. Unwin, 1887).

———— [1890], 'Dionea', in *Hauntings: The Supernatural Stories*, pp. 31–49.

———— [1890], *Hauntings*, 2nd edn (London: Lane, 1906).

———— [1890], 'The Legend of Madame Krasinka', in *Hauntings: The Supernatural Stories*, pp. 287–304.

———— [1890], 'A Wicked Voice', in *Hauntings: The Supernatural Stories*, pp. 87–105.

———— [1894], 'Ravenna and Her Ghosts', in *Pope Jacynth and More Supernatural Tales*, pp. 124–46.

———— [1909], *Laurus Nobilis: Chapters on Art and Life* (London: Lane, 1909).

———— [1913], 'The Gods and Ritter Tanhuser', in *Hauntings: The Supernatural Stories*, pp. 195–200.

———— [1927], 'The Doll', in *Hauntings: The Supernatural Stories*, pp. 277–83.

———— [1927], 'Introduction: For Maurice: Five Unlikely Stories', in *Hauntings: The Supernatural Stories*, pp. 177–91.

———— [1932], *Music and Its Lovers: An Empirical Study of Emotion and Imaginative Responses to Music* (London: G. Allen & Unwin, 1932).

———— *Pope Jacynth and More Supernatural Tales* (London: Owen, 1956).

———— *Hauntings: The Supernatural Stories* (Ashcroft: Ash-Tree Press, 2002).

Lee, Vernon and Clementina Anstruther-Thomson [1897], 'Beauty and Ugliness', in *Beauty and Ugliness and Other Studies on Psychological Aesthetics* (London: Lane, 1912), pp. 153–239.

Light Through the Crannies: Parables and Teachings from the Other Side (London: Longmans, 1888).

Macnish, Robert [1830], 'The Case of Mary Reynolds', in *Embodied Selves: An Anthology of Psychological Texts, 1830–1890*, ed. by Jenny Bourne Taylor and Sally Shuttleworth (Oxford: Clarendon Press, 1998), pp. 123–4.

Marsh, Richard [1897], *The Beetle* (Peterborough: Broadview, 2004).

Mitchell, S. Weir [1871], *Wear and Tear, or Hints for the Overworked*, 8th edn (Philadelphia: Lippincott, 1897).

Molesworth, Mary Louisa [1873], 'Lady Farquhar's Old Lady: A True Ghost Story', in *The Penguin Book of Classic Fantasy*, ed. by Susan A. Williams (London: Penguin, 1995), pp. 272–85.

———— [1888], *Four Ghost Stories* (London: Macmillan, 1888).

———— [1888], 'The Story of the Rippling Train', in *Four Ghost Stories*, pp. 227–55.

———— [1888], 'Unexplained', in *Four Ghost Stories*, pp. 87–226.

———— [1888], 'Witnessed by Two', in *Four Ghost Stories*, pp. 43–86.

Myers, A.T. [1886], 'The Life-History of a Case of Double or Multiple Personality', reprinted from *The Journal of Mental Science* (1886).

Myers, F.W.H. [1885], 'Automatic Writing or the Rationale of the Planchette', *The Contemporary Review*, 47 (1885), 233–49.

———— [1886], 'Multiplex Personality', in *Embodied Selves*, ed. by Jenny Bourne Taylor and Sally Shuttleworth, pp. 132–8.

———— [1903], *Human Personality and Its Survival of Bodily Death*, 2 vols (London: Longmans, 1903).

Nietzsche, Friedrich [1883–1885], *Thus Spoke Zarathustra*, trans. by R.J. Hollingdale (London: Penguin, 1969).

'Objects of the Society' [1882], in *The Fin de Siècle: A Reader in Cultural History, c. 1880–1900*, ed. by Sally Ledger and Roger Luckhurst (Oxford: Oxford University Press, 2000), pp. 271–2.

Oxley, William [1885], 'W.J. Colville's Lectures', in *The Medium and Daybreak* (London: J. Burns, 1885), vol. XVI, 4 September 1885, p. 567.

Paston, George [1898], *A Writer of Books* (Chicago: Academy Chicago Publishers, 1999).

Pater, Walter [1873], *The Renaissance* (Chicago: Pandora, c. 1977).

Podmore, Frank [1902], *The Rise of Victorian Spiritualism*, ed. by R.A. Gilbert, *Modern Spiritualism: A History and Criticism*, 8 vols (London: Routledge, 2000), vol. VII.

Poe, Edgar Allan [1845], 'The Facts in the Case of M. Valdemar', in *Tales of Mystery and Imagination*, intro. by John S. Whitley (Hertfordshire: Wordsworth, 1992), pp. 31–8.

Richet, Charles [1923], *Thirty Years of Psychical Research: Being a Treatise on Metaphysics*, trans. by Stanley de Brath (London: Collins, 1923).

Riddell, Charlotte [1882], 'Old Mrs. Jones', in *Weird Stories*, pp. 230–314.

——— [1882], 'The Open Door', in *Weird Stories*, pp. 48–103.

——— [1882], *Weird Stories*, new edn (London, 1885).

Schüle, Herbert [1892], 'Neuralgia', in *A Dictionary of Psychological Medicine*, ed. by D. Hack Tuke, II, pp. 835–40.

Segno, A. Victor [c. 1902], *The Law of Mentalism: A Practical, Scientific Explanation of Thought or Mind* (Los Angeles: American Institute of Mentalism Publishers, c. 1902).

'Spirit Photography', *Light*, 25 November 1893, p. 562.

Stead, W.T. [1891], *Real Ghost Stories: A Record of Authentic Apparitions* (London, 1891).

——— [1892], *More Ghost Stories: A Sequel to Real Ghost Stories* (London, 1892).

——— [1922], *The Blue Island: Experiences of a New Arrival Beyond the Veil* (London: Rider, 1922).

Stevenson, Robert Louis [1886], *Strange Case of Dr. Jekyll and Mr. Hyde*, ed. by Richard Drury (Edinburgh: Edinburgh University Press, 2004).

Stoker, Bram [1897], *Dracula* (Oxford: Oxford University Press, 1996).

Taylor, John Traill [1894], *The Veil Lifted: Modern Developments in Spirit Photography* (London: Whittaker & Co., 1894).

Townshend, Chauncy Hare [1840], *Facts in Mesmerism: with Reasons for a Dispassionate Inquiry into it*, 2nd edn (London: Baillière, 1844).

Tuke, D. Hack [1892], ed., *A Dictionary of Psychological Medicine*, 2 vols (London: Churchill, 1892).

Wilde, Oscar [1890–1891], *The Picture of Dorian Gray* (Harmondsworth: Penguin, 1970).

Secondary Sources

Anderson, Amanda and Joseph Valente, eds, *Disciplinarity at the Fin de Siècle* (Princeton: Princeton University Press, 2002).

Anger, Susy, *Victorian Interpretation* (Ithaca: Cornell University Press, 2005).

Archimedes, Sondra M., *Gendered Pathologies: The Female Body and Biomedical Discourse in the Nineteenth-Century English Novel* (London: Routledge, 2005).

Armstrong, Tim, *Modernism, Technology and the Body: A Cultural Study* (Cambridge: Cambridge University Press, 1998).

Auerbach, Nina, *Woman and the Demon: The Life of a Victorian Myth* (Cambridge: Cambridge University Press, 1982).

Barthes, Roland, *Camera Lucida: Reflections on Photography*, trans. by Richard Howard (London: Flamingo, 1984).

———, 'The Death of the Author', in *Authorship from Plato to the Postmodern: A Reader*, ed. by Séan Burke (Edinburgh: Edinburgh University Press, 1995), pp. 125–30.

Basham, Diana, *The Trial of Woman: Feminism and the Occult Sciences in Victorian Literature and Society* (London: Macmillan, 1992).

Batchen, Geoffrey, *Burning with Desire: The Conception of Photography* (Cambridge, MA: MIT Press, 1997).

Beizer, Janet, *Ventriloquized Bodies: Narratives of Hysteria in Nineteenth-Century France* (Ithaca: Cornell University Press, 1994).

Berthin, Christine, *Gothic Hauntings: Melancholy Crypts and Textual Ghosts* (Basingstoke: Palgrave Macmillan, 2010).

Bown, Nicola, Carolyn Burdett, and Pamela Thurschwell, 'Introduction', in *The Victorian Supernatural*, ed. by Nicola Bown, Carolyn Burdett, and Pamela Thurschwell, pp. 1–19.

———, eds, *The Victorian Supernatural* (Cambridge: Cambridge University Press, 2004).

Bradley, C. Alan and William A.S. Sarjeant, *Ms. Holmes of Baker Street: The Truth About Sherlock* (Alberta: University of Alberta Press, 2004).

Briggs, Julia, *Night Visitors: The Rise and Fall of the English Ghost Story* (London: Faber, 1977).

Bronfen, Elisabeth, *Over Her Dead Body: Death, Femininity and the Aesthetic* (Manchester: Manchester University Press, 1992).

Brookes, Martin, *Extreme Measures: The Dark Visions and Bright Ideas of Francis Galton* (New York: Bloomsbury, 2004).

Buse, Peter and Andrew Stott, 'Introduction', in *Ghosts: Deconstruction, Psychoanalysis, History*, ed. by Peter Buse and Andrew Stott (Basingstoke: Macmillan, 1998), pp. 1–20.

Caballero, Carlo, '"A Wicked Voice": On Vernon Lee, Wagner, and the Effects of Music', *Victorian Studies*, 35 (1992), 385–408.

Campbell, John L. and Trevor H. Hall, *Strange Things: The Story of Fr Allan McDonald, Ada Goodrich Freer, and the Society for Psychical Research's Enquiry into Highland Second Sight* (London: Routledge, 1968).

Castle, Terry, *The Apparitional Lesbian: Female Homosexuality and Modern Culture* (New York: Columbia University Press, 1993).

———, *The Female Thermometer: Eighteenth Century Culture and the Invention of the Uncanny* (Oxford: Oxford University Press, 1995).

Chapman, Alison, '"A Poet Never Sees a Ghost": Photography and Trance in Tennyson's *Enoch Arden* and Julia Margaret Cameron's Photography', *Victorian Poetry*, 41 (2003), 47–71.

Chapman, Alison and Jane Stabler, eds, *Unfolding the South: Nineteenth-Century British Women Writers and Artists in Italy* (Manchester: Manchester University Press, 2003).

Christ, Carol T. and John O. Jordan, eds, *Victorian Literature and the Victorian Visual Imagination* (Berkeley: University of California Press, 1995).

Christensen, Peter G., '"A Wicked Voice": Vernon Lee's Artist Parable', *Lamar Journal of the Humanities*, 15 (1989), 3–15.

Clark, Timothy, *The Theory of Inspiration: Composition as a Crisis of Subjectivity in Romantic and Post-Romantic Writing* (Manchester: Manchester University Press, 1997).

Coates, Paul, *The Double and the Other: Identity as Ideology in Post-Romantic Fiction* (New York: St. Martin's Press, 1988)

Colby, Vineta, *Vernon Lee: A Literary Biography* (Charlottesville: University of Virginia Press, 2003).

Connor, Steven, *Dumbstruck: A Cultural History of Ventriloquism* (Oxford: Oxford University Press, 2000).

Crabtree, Adam, *From Mesmer to Freud: Magnetic Sleep and the Roots of Psychological Healing* (New Haven: Yale University Press, 1993).

Crary, Jonathan, *Techniques of the Observer: On Vision and Modernity in the Nineteenth Century* (Cambridge, MA: MIT Press, 1998).

———, *Suspensions of Perception: Attention, Spectacle, and Modern Culture* (Cambridge, MA: MIT Press, 2001).

Daston, Lorraine and Peter Galison, 'The Image of Objectivity', *Representations*, 40 (1992), 81–128.

Davis, Michael, *George Eliot and Nineteenth-Century Psychology: Exploring the Unmapped Country* (Aldershot: Ashgate, 2006).

Day, William Patrick, *In the Circles of Fear and Desire: A Study of Gothic Fantasy* (Chicago: University of Chicago Press, 1985).

Dean, Bradley, *The Making of the Victorian Novelist: Anxieties of Authorship in the Mass Market* (London: Routledge, 2003).

Delamotte, Eugenia C., *Perils of the Night: A Feminist Study of Nineteenth-Century Gothic* (New York: Oxford University Press, 1990).

Dellamora, Richard, *Masculine Desire: The Sexual Politics of Victorian Aestheticism* (Chapel Hill: University of North Carolina Press, 1990).

———, ed., *Victorian Sexual Dissidence* (Chicago: University of Chicago Press, 1999).

Denisoff, Dennis, 'The Forest Beyond the Frame: Picturing Women's Desires in Vernon Lee and Virginia Woolf', in *Women and British Aestheticism*, ed. by Talia Schaffer and Kathy Alexis Psomiades (Charlottesville: University of Virginia Press, 1999), pp. 251–69.

————, '"Men of My Own Sex": Genius, Sexuality, and George Du Maurier's Artists', in *Victorian Sexual Dissidence*, ed. by Richard Dellamora (Chicago: University of Chicago Press, 1999), pp. 147–69.

Dickerson, Vanessa D., *Victorian Ghosts in the Noontide: Women Writers and the Supernatural* (Columbia: University of Missouri Press, 1996).

Dillingham, William B., 'Kipling: Spiritualism, Bereavement, Self-Revelation, and "They"', *ELT*, 45 (2002), 402–25.

Douglas-Fairhurst, Robert, *Victorian Afterlives: The Shaping of Influence in Nineteenth-Century Literature* (Oxford: Oxford University Press, 2002).

Dryden, Linda, *The Modern Gothic and Literary Doubles: Stevenson, Wilde and Wells* (Basingstoke: Palgrave Macmillan, 2003).

Edel, Leon, *Henry James: The Master 1901–1916* (London: Hart-Davis, 1972).

Ellis, Kate Ferguson, *The Contested Castle: Gothic Novels and the Subversion of Domestic Ideology* (Urbana: University of Illinois Press, 1989).

Ender, Evelyne, *Sexing the Mind: Nineteenth-Century Fictions of Hysteria* (Ithaca: Cornell University Press, 1995).

Farmer, Steve, 'Appendix F: Literary Censorship in Victorian England', in *The Story of a Modern Woman*, by Ella Hepworth Dixon (Peterborough: Broadview, 2004), pp. 267–88.

Ferguson, Christine, *Language, Science and Popular Fiction in the Victorian Fin-de-Siècle: The Brutal Tongue* (Aldershot: Ashgate, 2006).

Flint, Kate, *The Victorians and the Visual Imagination* (Cambridge: Cambridge University Press, 2000).

Forrest, Derek, *Hypnotism: A History* (London: Penguin, 1999).

Foucault, Michel, *The Archaeology of Knowledge*, trans. by A.M. Sheridan Smith (London: Tavistock, 1972).

Fraser, Hilary, *Beauty and Belief: Aesthetics and Religion in Victorian Literature* (Cambridge: Cambridge University Press, 1986).

Freud, Sigmund and Josef Breuer [1893–1895], *Studies on Hysteria*, ed. and trans. by James and Alix Strachey, *The Pelican Freud Library*, 15 vols (Harmondsworth: Penguin, 1974), vol. III.

————, [1912], 'The Dynamics of Transference', in *The Standard Edition of the Complete Psychological Works of Sigmund Freud*, XII, pp. 97–108.

————, 'The Uncanny' [1919], in *The Standard Edition of the Complete Psychological Works of Sigmund Freud*, XVII, pp. 219–56.

————, *The Standard Edition of the Complete Psychological Works of Sigmund Freud*, ed. and trans. by James Strachey, 24 vols (London: Hogarth Press, 1953–1974).

Frye, Lowell T., 'The Ghost Story and the Subjection of Women: The Example of Amelia Edwards, M.E. Braddon, and E. Nesbit', *Victorians Institute Journal*, 26 (1998), 167–209.

Fuller, Sophie and Nicky Losseff, eds, *The Idea of Music in Victorian Fiction* (Aldershot: Ashgate, 2004).

Galvan, Jill, *The Sympathetic Medium: Feminine Channeling, the Occult, and Communication Technologies, 1859–1919* (Ithaca: Cornell University Press, 2010).

Gauld, Alan, *A History of Hypnotism* (Cambridge: Cambridge University Press, 1992).

Gibson, John Michael and Richard Lancelyn Green, eds, *The Unknown Conan Doyle: Essays on Photography* (London: Secker & Warburg, 1982).

Gilder, Jeanette Leonard, *Trilbyana: The Rise and Progress of a Popular Novel* (New York: The Critic, 1895).

Gitelman, Lisa, *Scripts, Grooves, and Writing Machines: Representing Technology in the Edison Era* (Stanford: Stanford University Press, 1999).

Gordon, Lyndall, *A Private Life of Henry James: Two Women and His Art* (London: Norton, 1999).

Gracombe, Sarah, 'Converting Trilby: Du Maurier on Englishness, Jewishness, and Culture', *Nineteenth-Century Literature*, 58 (2003), 175–208.

Green-Lewis, Jennifer, *Framing the Victorians: Photography and the Culture of Realism* (Ithaca: Cornell University Press, 1996).

Griest, Guinevere L., *Mudie's Circulating Library and the Victorian Novel* (Newton Abbott: David & Charles, 1971).

Groth, Helen, *Victorian Photography and Literary Nostalgia* (Oxford: Oxford University Press, 2003).

Grove, Allen W., 'Röntgen's Ghosts: Photography, X-Rays, and the Victorian Imagination', *Literature and Medicine*, 16 (1997), 141–73.

Gunn, Peter, *Vernon Lee: Violet Paget, 1856–1935* (London: Oxford University Press, 1964).

Hacking, Ian, *Rewriting the Soul: Multiple Personality and the Sciences of Memory* (Princeton: Princeton University Press, 1995).

Hall, Jasmine Yong, 'Ordering the Sensational: Sherlock Holmes and the Female Gothic', *Studies in Short Fiction*, 28 (1991), 295–303.

Hall, Trevor H., *The Spiritualists: The Story of Florence Cook and William Crookes* (New York: Helix Press, 1963).

Harrington, Anne, *Medicine, Mind and the Double Brain: A Study in Nineteenth-Century Thought* (Princeton: Princeton University Press, 1987).

Harsh, Constance, 'Reviewing New Woman Fiction in the Daily Press: *The Times*, the *Scotsman*, and the *Daily Telegraph*', *Victorian Periodicals Review* 34 (2001), pp. 79–96.

Harvey, John, *Photography and Spirit* (London: Reaktion Books, 2007).

Heiland, Donna, *Gothic and Gender: An Introduction* (Oxford: Blackwell, 2004).

Heilmann, Ann, *New Woman Fiction: Women Writing First-Wave Feminism* (Basingstoke: Macmillan, 2000).

——, *New Woman Strategies: Sarah Grand, Olive Schreiner, Mona Caird* (Manchester: Manchester University Press, 2004).

Hogwood, Terence Allan and Kathryn Ledbetter, eds, *'Colour'd Shadows': Contexts in Publishing, Printing, and Reading Nineteenth-Century British Women Writers* (Basingstoke: Palgrave Macmillan, 2005).

Hughes, William, *Beyond Dracula: Bram Stoker's Fiction and Its Cultural Context* (Basingstoke: Palgrave Macmillan, 2000).

Hurley, Kelly, *The Gothic Body: Sexuality, Materialism and Degeneration at the Fin de Siècle* (Cambridge: Cambridge University Press, 1996).

Hutchison, Hazel, *Seeing and Believing: Henry James and the Spiritual World* (Basingstoke: Palgrave Macmillan, 2006).

Kali, Israel, *Names and Stories: Emilia Dilke and Victorian Culture* (Oxford: Oxford University Press, 2002).

Kaplan, Fred, *Henry James: The Imagination of Genius: A Biography* (London: Hodder & Stoughton, 1992).

Kelleher, Margaret, 'Charlotte Riddell's *A Struggle for Fame*: the Field of Women's Literary Production', *Colby Literary Quarterly*, 36 (2000), 116–32.

Kilgour, Maggie, *The Rise of the Gothic Novel* (London: Routledge, 1995).

Kissane, James and John M. Kissane, 'Sherlock Holmes and the Ritual of Reason', *Nineteenth-Century Fiction*, 17 (1963), 353–62.

Kittler, Friedrich, *Discourse Networks, 1800/1900*, trans. by Michael Metteer (Stanford: Stanford University Press, c. 1990).

Kontou, Tatiana, *Spiritualism and Women's Writing: From the Fin de Siècle to the Neo-Victorian* (Basingstoke: Palgrave Macmillan, 2009).

Kristeva, Julia, *Powers of Horror: An Essay on Abjection*, trans. by Leon S. Roudiez (New York: Columbia University Press, 1982).

Lancaster, Ashley Craig, 'Demonizing the Emerging Woman: Misrepresented Morality in *Dracula* and *God's Little Acre*', *Journal of Dracula Studies*, 6 (2004), 27–33.

Lansbury, Coral, 'Gynaecology, Pornography, and the Antivivisection Debate', *Victorian Studies*, 28 (1985), 413–37.

Lawler, Donald, 'The Gothic Wilde', in *Rediscovering Oscar Wilde*, ed. by C. George Sandulescu (Gerrards Cross: Smythe, 1994), pp. 249–68.

Lawrence, Frank, *Victorian Detective Fiction and the Nature of Evidence: The Scientific Investigations of Poe, Dickens, and Doyle* (Basingstoke: Palgrave Macmillan, 2003).

Ledger, Sally, *The New Woman: Fiction and Feminism at the Fin de Siècle* (Manchester: Manchester University Press, 1997).

Ledger, Sally and Roger Luckhurst, eds, *The Fin de Siècle: A Reader in Cultural History, c. 1880–1900* (Oxford: Oxford University Press, 2000).

Lehman, Amy, *Victorian Women and the Theatre of Trance: Mediums, Spiritualists and Mesmerists in Performance* (Jefferson: McFarland and Co., 2009).

Leighton, Angela, 'Ghosts, Aestheticism, and Vernon Lee', *Victorian Literature and Culture*, 28 (2000), 1–14.

———, 'Resurrections of the Body: Women Writers and the Idea of the Renaissance', in *Unfolding the South: Nineteenth-Century British Women Writers and Artists in Italy*, ed. by Alison Chapman and Jane Stabler (Manchester: Manchester University Press, 2003), pp. 222–38.

Leighton, Mary Elizabeth, 'Under the Influence: Crime and Hypnotic Fictions of the *Fin de Siècle*', in *Victorian Literary Mesmerism*, ed. by Martin Willis and Catherine Wynne (Amsterdam: Rodopi, 2006), pp. 203–226.

Liggins, Emma, 'Writing Against the "Husband-Fiend": Syphilis and Male Sexual Vice in the New Woman Novel', *Women's Writing*, 7 (2000), 175–95.

Logan, Peter Melville, *Nerves and Narratives: A Cultural History of Hysteria in Nineteenth-Century British Prose* (Berkeley: University of California Press, 1997).

London, Bette, *Writing Double: Women's Literary Partnerships* (Ithaca: Cornell University Press, 1999).

Luckhurst, Roger, *The Invention of Telepathy: 1870–1901* (Oxford: Oxford University Press, 2002).

Luckhurst, Roger, and Josephine McDonagh, eds, *Transactions and Encounters, Science and Culture in the Nineteenth Century* (Manchester: Manchester University Press, 2002).

MacLeod, Kirsten, *Fictions of British Decadence: High Art, Popular Writing and the Fin de Siècle* (Basingstoke: Palgrave Macmillan, 2006).

Maitland, Edward, *Anna Kingsford, Her Life, Letters, Diary, and Work*, 2nd edn (London, 1896).

Maltz, Diana, 'Engaging "Delicate Brains": From Working Class Enculturation to Upper-Class Lesbian Liberation in Vernon Lee and Kit Anstruther-Thomson's Psychological Aesthetics', in *Women and British Aestheticism*, ed. by Talia Schaffer and Kathy Psomiades (Charlottesville: University of Virginia Press, 1999), pp. 211–29.

Manocchi, Phyllis F., 'Vernon Lee and Kit Anstruther-Thomson: A Study of Love and Collaboration between Romantic Friends', *Women's Studies*, 12 (1986), 129–48.

Mansfield, Elizabeth, 'Emilia Dilke, Self-Fashioning and the Nineteenth Century', in *Marketing the Author: Authorial Personae, Narrative Selves and Self-Fashioning, 1880–1930*, ed. by Marysa Demoor (Basingstoke: Palgrave Macmillan, 2004), pp. 19–39.

Marien, Mary Warner, *Photography: A Cultural History*, 3rd edn (New York: Harry N. Abrams, 2002)

Matus, Jill, *Shock, Memory and the Unconscious in Victorian Fiction* (Cambridge: Cambridge University Press, 2009).

Maxwell, Catherine, 'From Dionysus to "Dionea": Vernon Lee's Portraits', *Word and Image*, 13 (1997), 253–69.

———, 'Vision and Visuality', in *A Companion to Victorian Poetry*, ed. by Richard Cronin, Alison Chapman, and Antony H. Harrison (Oxford: Blackwell, 2002), pp. 510–25.

———, 'Vernon Lee and the Ghosts of Italy', in *Unfolding the South*, ed. by Alison Chapman and Jane Stabler (Manchester: Manchester University Press, 2003), pp. 201–21.

———, *Second Sight: The Visionary Imagination in Late Victorian Literature* (Manchester: Manchester University Press, 2008).

Maxwell, Catherine and Patricia Pulham, eds, *Vernon Lee: Decadence, Ethics, Aesthetics* (Basingstoke: Palgrave Macmillan, 2006).

McDonald, Peter D., *British Literary Culture and Publishing Practice, 1880–1914* (Cambridge: Cambridge University Press, 1997).

McGarry, Molly, *Ghosts of Futures Past: Spiritualism and the Cultural Politics of Nineteenth-Century America* (Berkeley: University of California Press, 2008).

Meissner, Collin, '"What ghosts will be left to walk": Mercantile Culture and the Language of Art', *Henry James Review*, 21 (2000), 242–52.

Menke, Richard, *Telegraphic Realism: Victorian Fiction and Other Information Systems* (Stanford: Stanford University Press, 2008).

Mighall, Robert, *A Geography of Victorian Gothic Fiction: Mapping History's Nightmares* (Oxford: Oxford University Press, 1999).

Milbank, Alison, *Daughters of the House: Modes of the Gothic in Victorian Fiction* (Basingstoke: Macmillan, 1992).

Miles, Robert, 'Abjection, Nationalism and the Gothic', in *The Gothic*, ed. by Fred Botting (Woodbridge: Brewer, 2001), pp. 47–70.

Moody, Andrew J., '"The Harmless Pleasure of Knowing": Privacy in the Telegraph Office and Henry James's "In the Cage"', *Henry James Review*, 16 (1995), 53–65.

Morey, Peter, 'Gothic and Supernatural: Allegories at Work and Play in Kipling's Indian Fiction', in *The Victorian Gothic: Literary and Cultural Manifestations in the Nineteenth Century*, ed. by Ruth Robbins and Julian Wolfreys (Basingstoke: Palgrave, 2000), pp. 201–17.

Mulvey-Roberts, Marie, '*Dracula* and the Doctors: Bad Blood, Menstrual Taboo and the New Woman', in *Bram Stoker: History, Psychoanalysis and the Gothic*, ed. by William Hughes and Andrew Smith (Basingstoke: Macmillan, 1998).

Murphy, Patricia, 'Reevaluating Female "Inferiority": Sarah Grand versus Charles Darwin', *Victorian Literature and Culture*, 26 (1998), 221–36.

———, *In Science's Shadow: Literary Constructions of Late Victorian Women* (Columbia, Missouri: University of Missouri Press, 2006).

Noakes, Richard, 'Spiritualism, Science and the Supernatural in Mid-Victorian Britain', in *The Victorian Supernatural*, ed. by Nicola Bown, Carolyn Burdett, and Pamela Thurschwell (Cambridge: Cambridge University Press, 2004), pp. 23–43.

Novak, Daniel, *Realism, Photography, and Nineteenth-Century Fiction* (Cambridge: Cambridge University Press, 2008).

Oppenheim, Janet, *The Other World: Spiritualism and Psychical Research in England, 1850–1914* (Cambridge: Cambridge University Press, 1985).

Otis, Laura, *Networking: Communicating with Bodies and Machines in the Nineteenth Century* (Ann Arbor: University of Michigan Press, 2001).

Oulton, Carolyn W. de la L. and Sue Ann Schatz, eds, *Mary Cholmondeley Reconsidered*, Gender and Genre Series (London: Pickering & Chatto, 2010).

Owen, Alex, *The Darkened Room: Women, Power and Spiritualism in Late Victorian England* (London: Virago, 1989).

Pamboukian, Sylvia, 'Science, Magic and Fraud in the Short Stories of Rudyard Kipling', *ELT*, 47 (2004), 429–45.

Pearl, Sharrona, 'Dazed and Abused: Gender and Mesmerism in Wilkie Collins', in *Victorian Literary Mesmerism*, ed. by Martin Willis and Catherine Wynne (Amsterdam: Rodopi, 2006), pp. 163–82.

Petersen, Linda H, 'Charlotte Riddell's *A Struggle for Fame*: Myths of Authorship, Facts of the Market', *Women's Writing*, 11 (2004), 99–115.

Pick, Daniel, *Svengali's Web: The Alien Enchanter in Modern Culture* (London: Yale University Press, 2000).

Porter, Roy, Helen Nicholson, and Bridget Bennett, eds, *Women, Madness, and Spiritualism*, 2 vols (London: Routledge, 2003).

Price, Leah and Pamela Thurschwell, eds, *Literary Secretaries/Secretarial Culture* (Aldershot: Ashgate, 2005).

Psomiades, Kathy Alexis, *Beauty's Body: Femininity and Representation in British Aestheticism* (Stanford: Stanford University Press, 1997).

——, '"Still Burning from this Strangling Embrace": Vernon Lee on Desire and Aesthetics', in *Victorian Sexual Dissidence*, ed. by Richard Dellamora (Chicago: University of Chicago Press, 1999), pp. 21–41.

Punter, David, *The Literature of Terror: A History of Gothic Fiction from 1765 to the Present Day*, 2nd edn, 2 vols (London: Longman, 1996).

Pulham, Patricia, *Art and the Transitional Object in Vernon Lee's Supernatural Tales* (Aldershot: Ashgate, 2008).

Purcell, Edward L, 'Trilby and Trilby-Mania, The Beginning of the Bestseller System', *Journal of Popular Culture*, 11 (1977), 62–76.

Richardson, Angelique, 'Allopathic Pills? Health, Fitness and New Woman Fictions', *Women: A Cultural Review*, 10 (1999), 1–20.

Richardson, Angelique and Chris Willis, eds, *The New Woman in Fiction and Fact: Fin-de-siècle Feminisms* (Basingstoke: Palgrave Macmillan, 2002).

Robbins, Ruth and Julian Wolfreys, eds, *The Victorian Gothic: Literary and Cultural Manifestations in the Nineteenth Century* (Basingstoke: Palgrave, 2000).

Rowe, Katherine, *Dead Hands: Fictions of Agency, Renaissance to Modern* (Stanford: Stanford University Press, 1999).

Royle, Nicholas, *Telepathy and Literature: Essays on the Reading Mind* (Oxford: Blackwell, 1990).

——, *The Uncanny: An Introduction* (Manchester: Manchester University Press, 2003).

Ruddick, Nicholas, 'Life and Death by Electricity in 1890: The Transfiguration of William Kemmler', *Journal of American Culture*, 21 (1998), 79–87.

Ryan, James R., *Picturing Empire: Photography and the Visualization of the British Empire* (Chicago: University of Chicago Press, 1997).

Rylance, Rick, *Victorian Psychology and British Culture, 1850–1880* (Oxford: Oxford University Press, 2000).

Salmonsen, Jessica Amanda, ed., 'Preface', in *What Did Miss Darrington See? An Anthology of Feminist Supernatural Fiction* (New York: Feminist Press, 1989), pp. ix–xiv.

Savoy, Eric, '"In the Cage" and the Queer Effects of Gay History', *NOVEL*, 28 (1995), 284–307.

————, 'Spectres of Abjection: the Queer Subject of James's "The Jolly Corner"', in *Spectral Readings: Towards a Gothic Geography*, ed. by Glennis Byron and David Punter (Basingstoke: Macmillan, 1999), pp. 161–74.

Schaffer, Talia, *The Forgotten Female Aesthetes: Literary Culture in Late-Victorian England* (Charlottesville: University of Virginia Press, 2000).

Schaffer, Talia and Kathy Alexis Psomiades, eds, *Women and British Aestheticism* (Charlottesville: University of Virginia Press, 1999).

Schaper, Susan, 'Victorian Ghostbusting: Gendered Authority in the Middle-Class Home', *The Victorian Newsletter*, 100 (2001), 6–13.

Sekula, Allan, 'On the Invention of Photographic Meaning', in *Photography in Print: Writings from 1816 to the Present*, ed. by Vicki Goldberg (New York: Simon and Schuster, 1981), pp. 452–73.

Seltzer, Mark, 'The Postal Unconscious', *Henry James Review*, 21 (2000), 197–206.

Shattock, Joanne, ed., *Women and Literature in Britain, 1800–1900* (Cambridge: Cambridge University Press, 2001).

Shaw, Marion, '"To tell the truth of sex": Confession and Abjection in Late Victorian Writing', in *Rewriting the Victorians: Theory, History, and the Politics of Gender*, ed. by Linda M. Shires (New York: Routledge, 1992), pp. 87–100.

Showalter, Elaine, *The Female Malady: Women, Madness and English Culture, 1830–1980* (London: Virago, 1987).

————, *Sexual Anarchy: Gender and Culture at the Fin de Siècle* (London: Bloomsbury, 1991).

Shuttleworth, Sally, *Charlotte Brontë and Victorian Psychology* (Cambridge: Cambridge University Press, 1996).

Siebers, Alisha, 'Marie Corelli's Magnetic Revitalizing Power', in *Victorian Literary Mesmerism*, ed. by Martin Willis and Catherine Wynne (Amsterdam: Rodopi, 2006), pp. 183–202.

Simon, Linda, *Dark Light: Electricity and Anxiety from the Telegraph to the X-Ray* (London: Harcourt, 2004).

Smajic, Srdjan, *Ghost-Seers, Detectives, and Spiritualists: Theories of Vision in Victorian Literature and Science* (Cambridge: Cambridge University Press, 2010).

Smith, Andrew, *Victorian Demons: Medicine, Masculinity and the Gothic at the Fin-de-Siècle* (Manchester: Manchester University Press, 2004).

————, *Gothic Literature* (Edinburgh: Edinburgh University Press, 2007).

Smith, Lindsay, *The Politics of Focus: Women, Children and Nineteenth Century Photography* (Manchester: Manchester University Press, 1998).

Standage, Tom, *The Victorian Internet: The Remarkable Story of the Telegraph and the Nineteenth Century's Online Pioneers* (London: Weidenfeld and Nicolson, 1998).

Stashower, Daniel, *Teller of Tales: The Life of Arthur Conan Doyle* (London: Penguin, 2001).

Stetz, Margaret D., 'The Laugh of the New Woman', in *The Victorian Comic Spirit: New Perspectives*, ed. by Jennifer A. Wagner-Lawlor (Aldershot: Ashgate, 2000), pp. 219–41.

Stewart, Clare, '"Weird Fascination": The Response to Victorian Women's Ghost Stories', in *Feminist Readings of Victorian Popular Texts: Divergent Femininities*, ed. by Emma Liggins and Daniel Duffy (Aldershot: Ashgate, 2001), pp. 108–25.

Stiles, Anne, ed., *Neurology and Literature, 1860–1920* (Basingstoke: Palgrave Macmillan, 2007).

Suranyi, Clarissa J., 'Introduction', in *The Type-Writer Girl*, by Grant Allen (Peterborough: Broadview, 2004), pp. 9–17.

Sword, Helen, *Ghostwriting Modernism* (Ithaca: Cornell University Press, 2002).

Tagg, John, *The Burden of Representation: Essays on Photographies and Histories* (London: Macmillan, 1988).

Tatar, Maria M., *Spellbound: Studies on Mesmerism and Literature* (Princeton: Princeton University Press, 1978).

Taylor, Charles, *Varieties of Religion Today: William James Revisited* (Cambridge, MA: Harvard University Press, c. 2002).

Taylor, Jenny Bourne, *In the Secret Theatre of Home: Wilkie Collins, Sensation Narrative, and Nineteenth-Century Psychology* (London: Routledge, 1988).

Taylor, Jenny Bourne and Sally Shuttleworth, eds, *Embodied Selves: An Anthology of Psychological Texts, 1830–1890* (Oxford: Clarendon Press, 1998).

Taylor-Ide, Jesse Oak, 'Ritual and the Liminality of Sherlock Holmes in *The Sign of Four* and *The Hound of the Baskervilles*', *ELT*, 48 (2005), 55–70.

Thomas, Ronald R., 'The Fingerprint of the Foreigner: Colonizing the Criminal Body in 1890s Detective Fiction and Criminal Anthropology', *ELH*, 61 (1994), 655–83.

———, 'Making Darkness Visible: Capturing the Criminal and Observing the Law in Victorian Photography and Detective Fiction', in *Victorian Literature and the Victorian Visual Imagination*, ed. by Carol T. Christ and John O. Jordan (Berkeley: University of California Press, 1995), pp. 134–68.

Thurschwell, Pamela, *Literature, Technology and Magical Thinking, 1880–1920* (Cambridge: Cambridge University Press, 2001).

Tooley, Sarah A., 'The Woman's Question: An Interview with Madame Sarah Grand', in *Sex, Social Purity and Sarah Grand: Journalistic Writings and Contemporary Reception*, ed. by Ann Heilmann and Stephanie Forward, 4 vols (London: Routledge, 2000), I, pp. 220–29.

Trelease, Gita Panjabe, 'Time's Hand: Fingerprints, Empire, and Victorian Narratives of Crime', in *Victorian Crime, Madness and Sensation*, ed. by Andrew Maunder and Grace Moore (Aldershot: Ashgate, 2004), pp. 195–206.

Tromp, Marlene, *Altered States: Sex, Nation, Drugs, and Self-Transformation in Victorian Spiritualism* (Albany: State University of New York Press, 2006).

Tucker, Jennifer, 'Photography as Witness, Detective and Impostor: Visual Representations in Victorian Science', in *Victorian Science in Context*, ed. by Bernard Lightman (Chicago: University of Chicago Press, 1997), pp. 378–408.

van Schlun, Betsy, *Science and the Imagination: Mesmerism, Media and the Mind in Nineteenth-Century English and American Literature* (Berlin: Galda + Wilch Verlag, 2007).

Vrettos, Athena, 'Victorian Psychology', in *A Companion to the Victorian Novel*, ed. by Patrick Brantlinger and William B. Thesing (Oxford: Blackwell, 2002), pp. 67–83.

Wagner, Cosima, *Cosima Wagner's Diaries*, ed. and annotated by Martin Gregor-Dellin and Dietrich Mack, trans. and intro. by Geoffrey Skelton (London: Pimlico, 1974).

Walkowitz, Judith R., *Prostitution and Victorian Society: Women, Class, and the State* (Cambridge: Cambridge University Press, 1980).

———, *City of Dreadful Delight: Narratives of Sexual Danger in Late-Victorian London* (London: Virago, 1992).

Warne, Vanessa and Colette Colligan, 'The Man Who Wrote a New Woman Novel: Grant Allen's *The Woman Who Did* and the Gendering of New Woman Authorship', *Victorian Literature and Culture*, 33 (2005), 21–46.

Warner, Marina, *Phantasmagoria: Spirit Visions, Metaphors and Media into the Twenty-First Century* (Oxford: Oxford University Press, 2006).

Weliver, Phyllis, *Women Musicians in Victorian Fiction, 1860–1900: Representations of Music, Science, and Gender in the Leisured Home* (Aldershot: Ashgate, 2000).

Wilburn, Sarah A., *Possessed Victorians: Extra Spheres in Nineteenth-Century Mystical Writings* (Aldershot: Ashgate, 2006).

Williams, Erin, 'Female Celibacy in the Fiction of Gissing and Dixon: The Silent Strike of the Suburbanites', *ELT*, 45 (2002), 259–79.

Willis, Martin, *Mesmerists, Monsters and Machines: Science Fiction and Cultures of Science in the Nineteenth Century* (Kent, Ohio: Kent State University Press, 2006).

Willis, Martin and Catherine Wynne, eds, *Victorian Literary Mesmerism* (Amsterdam: Rodopi, 2006).

Winter, Alison, *Mesmerized: Powers of Mind in Victorian Britain* (Chicago: University of Chicago Press, 1998).

Wolfreys, Julian, *Victorian Hauntings: Spectrality, Gothic, the Uncanny and Literature* (Basingstoke: Palgrave Macmillan, 2002).

Wolstenholme, Susan, 'Possession and Personality: Spiritualism in *The Bostonians*', *American Literature: A Journal of Literary History, Criticism, and Bibliography*, 49 (1978), 580–91.

Wood, Jane, *Passion and Pathology in Victorian Fiction* (Oxford: Oxford University Press, 2001).

Wynne, Catherine, *The Colonial Conan Doyle: British Imperialism, Irish Nationalism, and the Gothic* (London: Greenwood, 2002).

Zorn, Christa, *Vernon Lee: Aesthetics, History and the Victorian Female Intellectual* (Athens: Ohio University Press, 2003).

Index

abjection (*see also* Kristeva, Julia) 87,
    101–9, 119
aestheticism (*see* Gothic, the; Lee, Vernon)
afflatus 31n57, 162
Allen, Grant
    Type-Writer Girl, The 11, 139, 149–52
altered states of perception (*see also*
        extrasensory perception; trance)
        10, 81, 87, 93, 95
animal magnetism 62, 72
Anstruther-Thomson, Clementina (Kit)
        122–4, 126–7
Archimedes, Sondra M. 94
authorship 3, 9, 15n5, 17–8, 25, 26, 28, 76,
        87, 88, 95n47, 100, 116, 139, 151,
        152n27, 157, 158, 161, 162
automatic writing (*see also* discursive
        technologies; medium, spiritualist;
        telegraphy; typewriter) 7, 8, 11, 16,
        18, 27–35, 87n10, 100, 126, 139,
        144–5, 150–51, 153

Baldwin, Louisa 91
    'Many Waters Cannot Quench
        Love' 97
Ballechin house 86, 87–90
Barter, John 67–9, 71–2
Barthes, Roland 60
Basham, Diana 5n13, 83n3, 84n6,
        86n8, 91n28
Batchen, Geoffrey 46, 60
Beizer, Janet 94
Berenson, Bernard 123
Berthin, Christine 2
Bertillon system 49
Besant, Annie 99–100
Blavatsky, Helena Petrovna 91, 95, 97, 98,
        99–101
    'A Bewitched Life (As Narrated by a
        Quill Pen)' 100
    'The Cave of the Echoes' 100

'The Ensouled Violin' 100
    Nightmare Tales 99–100
Bosanquet, Theodora 9, 26–8, 34, 36
Bourru, Hippolyte 20n19, 21–2
Bown, Nicola 7
Braid, James 63, 65, 72n51, 74–5, 77
Briggs, Julia 91
Burdett, Carolyn 7
Burot, Prosper 20n19, 21–2
Buse, Peter 119

Caballero, Carlo 119n28, 121n32, 132n61
Campbell, John L. 88–9, 90n23
Carpenter, William Benjamin 7, 62,
        74, 128
Castle, Terry 29, 120n29
celebrity 9,17, 24, 28, 76, 158
Charcot, Jean Martin 63, 65n19, 67n25,
        69, 70, 72
Cholmondeley, Mary
    Red Pottage 145, 150
Christensen, Peter G. 120n28, 132n61
clairvoyance 33n63, 68–9, 70, 87n10
Clark, Timothy 143
Coates, James 67–70, 73–5
Coates, Mrs. James 54n93
Coates, Paul 19
Cobban, J. MacLaren
    Master of His Fate, The 72n52, 80
Colby, Vineta 120–21, 124
Colligan, Colette 151n27
Collins, Wilkie 79n68
    Moonstone, The 143n10
composite portraits or composite
        photographs 9, 49–53, 64, 89
    Sherlock Holmes stories 51–2
Connor, Steven 16n10, 30n54
consciousness 20–21, 84, 85–6, 94, 98,
        100n61, 144–5, 146
Cook, Florence (*see also* King, Katie) 40,
        126–7

Corelli, Marie
  Romance of Two Worlds, A 139n6
  Ziska 80
Crary, Jonathan 38n8, 40, 60
Crookes, William 7, 44n34, 89n17,
    126–7, 134

Davis, Michael 20
Denisoff, Dennis 79n68, 120n28
Dickens, Charles 4, 5, 62, 66n24, 77
Dickerson, Vanessa D. 86, 91n28, 102
Dilke, Emilia 91, 95, 101, 103–6
  'The Black Veil' 105
  'The Physician's Wife' 104–5
  'The Shrine of Death' 103–4, 105–6
  *Shrine of Death and Other Stories,
    The* 105
  *Shrine of Love and Other Stories,
    The* 105
  'The Silver Cage' 104
discursive technologies (*see also* automatic
    writing; telegraphy; typewriter) 5,
    16, 28–30, 33
Dixon, Ella Hepworth
  Story of a Modern Woman, The 138n2,
    147, 148n20, 159
double consciousness 5, 20n19, 94, 146
double personality 146
Douglas-Fairhurst, Robert 16n7
Doyle, Arthur Conan (*see also* genre;
    hypnotism; medium, spiritualist;
    mesmerism; photography; spirit
    photography; spiritualism; trance)
    1, 2, 3, 5, 6, 9, 11, 42, 84
  'The Adventure of the Cardboard Box'
    37–8, 59–60, 153n30
  'The Adventure of the Final Problem'
    37n2, 48, 60
  'The Adventure of the Illustrious
    Client' 58, 60
  'The Adventure of the Naval Treaty' 49
  'The Adventure of the Red Circle' 58–9
  'The Adventure of the Shoscombe Old
    Place' 54
  'The Adventure of the Speckled Band'
    49n57, 49n58, 55n94, 56n105
  'The Adventure of the Sussex
    Vampire' 54
  'The Adventure of the Veiled Lodger' 43

  'After Cormorants with a Camera' 43, 45
  'The Boscombe Valley Mystery'
    49n57, 55, 56
  Case for Spirit Photography, The 47, 83
  Coming of the Fairies, The 47
  'The Crooked Man' 57
  'A Few Technical Hints' 43n28
  History of Spiritualism, The 40–41,
    47–8
  Hound of the Baskervilles, The 49n57,
    49n60, 51–2, 54, 57n110, 60
  'The Man with the Twisted Lip' 44
  'On the Slave Coast with a Camera' 45–6
  Parasite, The 59, 66, 80
  'A Scandal in Bohemia' 38n5, 53–5, 58
  Sign of Four, The 44–5, 47n50, 49n57,
    55n96, 58n115
  Study in Scarlet, A 38n5, 47n50,
    49n57, 55–6, 57n108, 59n120
  'Trial of Burton's Emulsion Process'
    43n28
  Vital Message, The 44, 45n36, 55n93
dream states 2, 10, 87, 95, 98
Du Cane, Edmund 50, 51
Du Maurier, George (*see also* hypnotism;
    influence; mesmerism; trance) 6, 76
  Trilby 1, 10, 61–2, 64–6, 72–81, 129,
    131, 133

ecstasy (*see* trance)
ectoplasm (*see* imponderable fluids)
Edel, Leon 15n4, 26–7, 28n47, 36
electricity 29, 30, 37, 62, 72, 126–7, 161
electromagnetism 60, 72n51
extrasensory perception (*see also* altered
    states of perception; trance) 1,
    138, 140, 142, 143n10, 145,
    146, 148, 160

Faraday, Michael 72n51
fingerprinting 42, 49
Flint, Kate 38
Freer, Miss (*see* Goodrich-Freer, Ada)
Freud, Sigmund 92n29, 145
  hypnotism 63n10, 70n41, 94
  hysteria 70n41, 94
  transference 70
  uncanny 7–9, 17, 19, 131
Fullerton, W. Morton 24–5, 30

Galbraith, Lettice 91
'The Ghost in the Chair' 97
Galton, Francis 9, 42, 49–53, 64, 89
Galvan, Jill 5, 34, 36n69, 152
Gauld, Alan 61, 63, 69
gender 3, 5, 26, 27n45, 62n8, 66, 73, 79, 81,
85, 86, 90, 93, 98, 100n61, 112n2,
115n11, 131, 132n61, 147n20
genre 6
detective fiction 39, 45n38
ghost story 5, 91, 98, 100, 106
the Gothic 1–2, 4
James, Henry 115–8
Lee, Vernon 10, 111, 112, 115–8,
120n28, 121, 129
New Woman fiction 138
ghost-seeing 83, 87, 90, 95, 106
ghost story (*see* genre)
ghosts 2, 8, 15, 25, 39n12, 40, 41, 49, 51,
52, 53, 54, 57, 88, 161, 162
as inspirational and political symbols
in women's writing 10, 81, 83–109,
138, 162
Lee, Vernon 113, 117–8, 119, 121,
125–9, 130, 132, 133, 134, 161
questions of agency 16, 18, 24, 162
spiritualists' and mental scientists'
anxieties about materialism and 4,
10, 125–6, 127
theories of the haunted mind 6, 22, 69,
83–6, 87, 90, 113, 127–8
Gilman, Charlotte Perkins
'The Yellow Wallpaper' 92
Gissing, George
Odd Women, The 150
Gitelman, Lisa 17, 27, 34
Goodrich-Freer, Ada 87–90
Gordon, Lyndall 23
Gothic, the (*see also* genre)
abjection 101
aestheticism 112–13, 115n11
definition of the late Victorian Gothic
1–6, 10, 161
female Gothic 39n11, 90
mental science and 14–16, 48
New Woman fiction 137, 138,
141–2, 143
trance 66, 80

Gracombe, Sarah 78n66
Grand, Sarah 138
Beth Book, The 1, 10, 137n2, 139,
140–49, 152–3, 154, 156
Heavenly Twins, The 140n7
Green-Lewis, Jennifer 49
Gunn, Peter 121n32, 123
Gurney, Edmund 22, 84, 129–30

Hacking, Ian 20–21, 22
Hall, Jasmine Yong 39n11
Hall, Trevor H. 88–9
Hardy, Thomas 134
Far From the Madding Crowd 19
'The Fiddler of the Reels' 146n17
Harvey, John 44
haunted house 10, 15, 24, 90, 98–9, 106
Hawthorne, Nathaniel
Blithedale Romance, The 25n37
heightened perception 3, 9, 10, 34, 59,
121n32, 140, 143, 145, 146, 148,
152, 153, 161
Heilmann, Ann 138, 141n8, 154, 158
Hoffmann, E.T.A. 131
Hurley, Kelly 3, 4, 101
Huysmans, Joris-Karl
Against Nature 143n11
hypnotism (*see also* Du Maurier, George,
Trilby; influence; mesmerism;
Society for Psychical
Research; trance)
attempts to make distinct from
mesmerism 9, 59, 61, 63–4, 65,
66–70, 73, 74, 161
celebrity 75–6
Freud, Sigmund 63n10, 70n41, 94
hysteria 63, 64, 65, 69
music 146
photography 60
Sherlock Holmes 39, 58–60
sites of power 9–10, 61, 65,
66, 70–81
spiritualism 54n91
Stoker, Bram, *Dracula* 149–50
theories of the haunted mind 85, 86n8
transference 68, 70–72
hysteria 2, 10, 63, 64, 65, 69, 70n41, 86,
87, 91, 93–5, 106, 107, 160

imponderable fluids 2, 41, 59
    ectoplasm 2, 22, 30, 40–41, 47, 59
    mesmeric fluid (also known as
        magnetic or universal fluid) 3, 41,
        59, 61–2, 63, 65, 66, 69n39
    odic force 41
    psychic fluid 41
    spirit photography 41
    X-rays 40n14
influence
    anxieties about the dangers of influence
        in writing 16, 26, 28, 36, 45,
        100n61, 116, 140, 141, 153, 162
    dangers of foreign influence 66, 71, 80
    Du Maurier, George, *Trilby* 76, 78n66,
        79, 80
    hypnotism 54, 58, 66n21, 68, 75
    mesmerism 29, 62, 66, 129
    spiritual and supernatural 55, 56, 57,
        148, 158
intuition 55–6

Jacobs, Joseph 51
James, Henry (*see also* genre; trance) 3, 5,
        6, 11, 13, 14, 15–17, 20, 24–8, 36,
        48, 111, 128, 161
    Bostonians, The 80, 149n21
    'In the Cage' 1, 9, 17, 28, 33–6, 37,
        149, 150, 152
    'The Jolly Corner' 101n65
    'The Private Life' 1, 9, 17–20,
        23–4, 25
    Wings of the Dove, The 116
James, William 13–14, 15, 21–2, 23, 48,
        84, 89n17, 118, 121
Janet, Pierre 66–7, 77n64

Keats, John 30, 31, 32
    'The Eve of St. Agnes' 31, 32
    'This Living Hand' 32
Kellogg, John Harvey 92n29
King, Katie (*see also* Cook, Florence) 40,
        126–7
Kingsford, Anna Bonus 5, 91, 95, 97, 98,
        100, 101
    Dreams and Dream-Stories 98
    'Steepside' 98–9
Kipling, Rudyard 33n62, 33n63
    'Wireless' 1, 9, 17, 28–33, 34, 58

Kittler, Friedrich 17
Kontou, Tatiana 16n7, 100n61
Kristeva, Julia (*see also* abjection) 87,
        101–3, 105, 108–9
Külpe, Oswald 121–2

Lansbury, Coral 142n8
Lawler, Donald 113
Lawrence, D.H.
    Kangaroo 1, 7n22, 8n26, 161
Lee, Vernon (*see also* genre; ghosts;
        materialism; spiritualism;
        supernatural, the) 1, 6, 10, 11, 15,
        16, 161
    'Amour Dure' 112, 113, 116, 128
    'Beauty and Ugliness' 112, 121–4,
        127, 129
    Beauty and Ugliness 115n15, 121–2,
        123–4
    'Chapelmaster Kreisler' 131, 135n70
    'A Culture Ghost; or, Winthrop's
        Adventure' 128, 130, 131
    'Dionea' 112, 113–14
    'The Doll' 113, 114
    'Faustus and Helena: Notes on the
        Supernatural in Art' 111, 117, 119,
        125, 134
    'The Gods and Ritter Tanhuser' 117
    Hauntings 116n16, 125, 128n49, 130
    'The Hidden Door' 128–9
    'Impersonality and Evolution
        in Music' 129, 130n56, 133
    Juvenilia 115n15, 124
    'Lady Tal' 116
    'The Legend of Madame Krasinka'
        113, 114
    Miss Brown 116, 120n28
    Music and its Lovers 112, 123, 130
    'Oke of Okehurst; or, The Phantom
        Lover' 113, 128
    *Pope Jacynth and More Supernatural
        Tales* 118
    'Ravenna and Her Ghosts' 118–19
    *Studies of the Eighteenth Century in
        Italy* 121, 129
    'A Wicked Voice' 115, 119n28, 128,
        130–33
Lehman, Amy 56
Leighton, Angela 114–15, 134

Leighton, Mary Elizabeth 64, 65
Lind, Jenny 74–5, 76
Lipps, Theodor 123, 124
literary marketplace 18, 24, 28, 139, 156, 158, 159
Lodge, Oliver 40n15, 89n17
London, Bette 15n5, 35
Lowe, Louisa 93
Luckhurst, Roger 2–3, 5, 6, 14n3, 29n51, 30, 31n58, 66, 139

MacAlpine, William 26–7
Maitland, Edward 98
Maltz, Diana 124
Marsh, Richard 4
    Beetle, The 14n2, 66, 80, 137
materialism
    Lee, Vernon 10, 114–15, 125–6, 127, 132, 133, 134
    spiritualists' and mental scientists' anxieties about science, the supernatural and 4, 10, 22, 64, 125–6, 127, 161
Matus, Jill 21
Maxwell, Catherine 112, 116, 118n24, 120n28, 121n32, 134–5
McDonald, Fr Allan 88
McDougall, William 22
McGarry, Molly 94–5
medium, spiritualist (*see also* automatic writing) 3, 5, 8, 9, 16, 18, 22, 25, 27, 29, 30, 31, 34, 35, 36n69, 40, 44, 90, 100, 125, 126, 133, 144, 151, 152
    hysteria 93–5
    Sherlock Holmes as 54, 56–8
Meissner, Colin 24n34
Menke, Richard 38n6
mental science (*see* consciousness; double consciousness; Gothic, the; hypnotism; mesmerism; psychology; Society for Psychical Research; subconscious, the; subliminal consciousness; unconscious, the)
Mesmer, Franz Anton 61–2, 65, 66, 72, 156

mesmerism (*see also* Du Maurier, George, *Trilby*; hypnotism; imponderable fluids; influence; Society for Psychical Research; trance) 1, 2, 5, 29, 41, 61–2
    attempts to make distinct from hypnotism 9, 11, 59, 63–70, 161
    celebrity 75–6
    Doyle, Arthur Conan, *Parasite* 59
    music 129, 133, 146, 156–7
    photography 60
    sites of power 9, 10, 59, 61, 65, 66, 70–73, 75, 76, 78, 79n68, 80, 81, 149n21
    transference 68, 70–72
Mighall, Robert 3, 4, 14n3
Mitchell, S. Weir 92n29
Molesworth, Mary Louisa 5, 91, 95, 96, 101, 103
    'Lady Farquhar's Old Lady: A True Ghost Story' 96–7
    'The Story of the Rippling Train' 97
    'Unexplained' 106, 107–8
    'Witnessed by Two' 84–5
Moody, Andrew J. 35–6
Morse code 16, 29, 31
Mudie's circulating library 158
multiple (or multiplex) personality 20–22, 23, 77n64
Mumler, William 40
Murphy, Patricia 56, 147n20
music 65, 74, 115, 117, 121, 123, 128, 129–33, 135, 146, 148, 156–7
    the voice 16, 65, 74, 75, 77, 78, 115, 129–34
Myers, F.W.H. 13, 14, 15, 16, 21, 22, 29, 30, 67, 69, 77n64, 84, 85, 87, 88, 89, 98, 144

neuralgia 64, 72–3
neurosis 72, 94
New Woman 1, 2, 3, 6, 10–11, 59, 91, 92, 96, 137, 138–40, 141, 145, 146, 147n20, 149–50, 152, 153, 154, 157, 158, 160
Nietzsche, Friedrich 152
Noakes, Richard 7
Novak, Daniel 42, 51, 52

Oppenheim, Janet 16n6, 16n8, 41, 53, 57, 59, 127
Otis, Laura 30
Owen, Alex 16n6, 35, 93, 133, 134n63
Owen, Peter 118

Pamboukian, Sylvia 33n62
Paston, George
    Writer of Books, A 1, 10, 137, 138, 139, 140, 145, 146, 153–160
Pater, Walter 112, 116, 129
Pearl, Sharrona 79n68
phantasm 84
phonograph 16
photography (*see also* composite portraits; spirit photography) 9, 38–9, 41–2, 43, 45, 46, 49, 55, 60
Pick, Daniel 62, 65, 66, 74, 75, 76
Podmore, Frank 22, 84, 89–90, 126
Poe, Edgar Allan
    'The Facts in the Case of M. Valdemar' 66n24, 113
Psomiades, Kathy Alexis 115n11, 120n28, 124
psychical research (*see* Society for Psychical Research; spiritualism)
psychology 4, 13, 14n3, 20, 22, 42, 63, 67n26, 69, 72, 84n6, 112, 121, 123, 128
Pulham, Patricia 111n2, 132n61
Purcell, Edward L. 75

Reichenbach, Karl von 41
rest cure, the 92
Reynolds, Mary 20n19
Richet, Charles 22, 40, 69, 84, 89n17, 126
Riddell, Charlotte 91, 95, 101, 103
    'Old Mrs. Jones' 106
    'The Open Door' 95
Rowe, Katherine 25n38, 32n60
Royle, Nicholas 6, 8, 29n51
Ryan, James R. 41–2

Salmonsen, Jessica Amanda 91
Savoy, Eric 33n65, 101
scenes of writing (*see also* automatic writing; discursive technologies) 3, 4, 5n13, 6, 9, 11, 16, 17, 28, 116, 138, 139, 142, 161

intoxicating nature of 1, 3, 139, 142, 144, 147, 153
Schaffer, Talia 113
Schaper, Susan 106
séance 1, 5, 7, 16, 18, 22, 24, 25, 27, 30, 32, 33, 35, 40, 44, 48, 52, 53, 55, 57, 62, 69, 92n13, 93, 125, 126, 133, 134, 144, 153, 157n34
    as medical term 69
secondary self 16, 85, 144–5
Sekula, Allan 42n24
Seltzer, Mark 24n36
sensation fiction 143
Showalter, Elaine 93–4
Shuttleworth, Sally 20, 22
Siebers, Alisha 139n6
Smajic, Srdjan 39n12, 45n38, 84n5
Smith, Andrew 2n4, 3, 4
Smith, Lindsay 46
Society for Psychical Research (SPR) 2, 13, 33n63, 41, 89n17, 115, 125, 128
    hypnotism and mesmerism 69–70
    the investigation of the haunting of Ballechin house and treatment of Miss Freer 10, 86, 87–90
    theories of the haunted mind 13, 16, 22, 69–70, 86, 127, 144
    use of empirical science to test spiritualist phenomena 10, 47, 125–7
Society for the Study of Supernormal Pictures (SSSP) 46–7
somnambulism 69n39, 75, 98, 106
spirit photography (*see also* composite portraits; photography) 9, 39–41, 43, 44, 45, 46–7, 52, 53, 54n93, 57, 83
spiritualism 1, 2, 5, 6, 9, 7, 11, 16, 57, 98, 99, 133, 161
    concerns about materialism 4, 10, 125–6, 127
    Doyle, Arthur Conan 6, 9, 39, 40, 43–8, 60
    as empowering for women 92–3
    full-form materializations 7, 40n17, 126
    Galton, Francis 52–3
    hysteria 91, 94–5
    Kipling, Rudyard 33n63
    Lee, Vernon 10, 111, 115

light 57, 57n111
mental science 81
mesmerism 59
as pathological 93–4
periodicals: *Borderland* 83, 87n10;
    Light 41; *Medium and Daybreak,*
    The 56; *Spiritualist, The* 40
psychical research on 16, 22, 42,
    125–7, 128
Sherlock Holmes 54–8
spirit-rappings 1, 7, 41
'veil' 25, 25n37, 58
Stead, W.T. 6, 25n37, 83, 98
    Real Ghost Stories 10, 83–6, 87
Stevenson, Robert Louis 5
    Strange Case of Dr Jekyll and Mr
        Hyde 14, 21, 101n65
Stoker, Bram 5
    Dracula 14, 66, 77, 80, 137, 147n18,
        149–50, 152
Stott, Andrew 119
subconscious, the 16, 86, 124, 145
subliminal consciousness 14, 83
subliminal self 77
supernatural, the
    definitions of 6–7
        Lee, Vernon 111–12
        Victorian understanding of 7, 112
Suranyi, Clarissa J. 151
Sword, Helen 16n7

Taylor, Jenny Bourne 22
Taylor-Ide, Jesse Oak 39n11
telegraphy (*see also* automatic writing;
        discursive technologies) 3, 16, 17,
        27, 29–30, 30–31, 33–5, 37, 38n6,
        149, 152–3, 156
telepathy 1, 2, 3, 5, 6, 8, 22, 25, 29–32, 33,
        34, 35, 37, 56n101, 59, 68, 69, 84,
        129, 138, 139, 140, 141, 143n10,
        144, 145, 150, 152
telephone 16, 27, 30n54, 34
Thomas, Ronald R. 38–9, 49n61
Thurschwell, Pamela 5, 6, 7, 28
Townshend, Chauncy Hare 66
trance (*see also* altered states of
        perception; extrasensory
        perception; Gothic, the) 2, 49

Doyle, Arthur Conan, *Parasite* 59,
    66, 80
Du Maurier, George, *Trilby* 66, 76, 80
ecstasy 3, 139, 143, 144, 153
hypnotic 9, 39, 58–9, 63, 64, 65, 66,
    67, 68, 69, 71, 73, 76–7, 80, 81, 86,
    87, 106
inspiration as 143n11
James, Henry, *Bostonians* 149n21
Marsh, Richard, *Beetle* 66, 80
mesmeric 2, 9, 62, 64, 65, 66, 67,
    69n39, 70, 80, 81, 139, 143n10,
    144, 149, 156, 157
New Woman fiction 139, 152
Sherlock Holmes 39, 49, 58–60
spiritualist 16, 58, 86
Stoker, Bram, *Dracula* 66, 80
women's writing 10, 59
Tromp, Marlene 5, 35, 92, 93n32
Tucker, Jennifer 42
Tuke, Daniel Hack 64, 72
typewriter (*see also* automatic writing;
        discursive technologies) 3, 9, 11,
        16, 17, 24–8, 29, 30, 34, 150–51

uncanny, the
    definitions of 6–9
        Freud, Sigmund 7–9, 17, 19, 131
unconscious, the 10, 20–21, 29, 85–7, 98,
        128, 144–5, 146, 152
unconscious cerebration 16

van Schlun, Betsy 61n2, 62
venereal disease 71, 140n7
Vivet, Louis 21
vivisection 141n8, 142
Vrettos, Athena 62

Wagner, Cosima 1, 57
Walker, William 83
Warne, Vanessa 152n27
Weld, Mary 26, 27n45
Weldon, Georgina 93
Weliver, Phyllis 65, 146, 156, 157n34
Wilburn, Sarah A. 92n31, 157n34
Wilde, Oscar 5, 115, 116
    Picture of Dorian Gray, The
        14, 137, 142n9

Williams, Erin 147n20
Willis, Martin 62n8, 64
Winter, Alison 61n2, 62, 65, 69, 72n51, 80
Wolfreys, Julian 1–2, 4
Wolstenholme, Susan 149n21
Wynne, Catherine 46n44, 62n8

X., Felida 20n19
X., Miss (*see* Goodrich-Freer, Ada)
X-rays (*see* imponderable fluids)

Zorn, Christa 120n28